ASCE STANDARD

ASCE/SEI

43-19

AF768814

Seismic Design Criteria for Structures, Systems, and Components in Nuclear Facilities

AMERICAN SOCIETY OF CIVIL ENGINEERS

STRUCTURAL ENGINEERING INSTITUTE

PUBLISHED BY THE AMERICAN SOCIETY OF CIVIL ENGINEERS

Library of Congress Cataloging-in-Publication Data

Names: American Society of Civil Engineers, author.
Title: ASCE standard, ASCE/SEI, 43-19 : seismic design criteria for structures, systems, and components in nuclear facilities / American Society of Civil Engineers.
Other titles: Seismic design criteria for structures, systems, and components in nuclear facilities
Description: Reston, Virginia : American Society of Civil Engineers, [2019] | Includes bibliographical references and index. | Summary: "Standard ASCE/SEI 43-19 provides stringent criteria to ensure that nuclear facilities are designed to withstand the effects of earthquake ground shaking"– Provided by publisher.
Identifiers: LCCN 2019054308 | ISBN 9780784415405 (paperback) | ISBN 9780784482407 (pdf)
Subjects: LCSH: Nuclear facilities–Protection–Standards–United States. | Nuclear facilities–Earthquake effects. | Earthquake resistant design–Standards–United States.
Classification: LCC TK9152.163 .A45 2019 | DDC 621.48/32–dc23
LC record available at https://lccn.loc.gov/2019054308

Published by American Society of Civil Engineers
1801 Alexander Bell Drive
Reston, Virginia, 20191-4382
www.asce.org/bookstore | ascelibrary.org

This standard was developed by a consensus standards development process that has been accredited by the American National Standards Institute (ANSI). Accreditation by ANSI, a voluntary accreditation body representing public and private sector standards development organizations in the United States and abroad, signifies that the standards development process used by ASCE has met the ANSI requirements for openness, balance, consensus, and due process.

While ASCE's process is designed to promote standards that reflect a fair and reasoned consensus among all interested participants, while preserving the public health, safety, and welfare that is paramount to its mission, it has not made an independent assessment of and does not warrant the accuracy, completeness, suitability, or utility of any information, apparatus, product, or process discussed herein. ASCE does not intend, nor should anyone interpret, ASCE's standards to replace the sound judgment of a competent professional, having knowledge and experience in the appropriate field(s) of practice, nor to substitute for the standard of care required of such professionals in interpreting and applying the contents of this standard.

ASCE has no authority to enforce compliance with its standards and does not undertake to certify products for compliance or to render any professional services to any person or entity.

ASCE disclaims any and all liability for any personal injury, property damage, financial loss, or other damages of any nature whatsoever, including without limitation any direct, indirect, special, exemplary, or consequential damages, resulting from any person's use of, or reliance on, this standard. Any individual who relies on this standard assumes full responsibility for such use.

***Errata:* Errata, if any, can be found at http://dx.doi.org/10.1061/9780784415405**

ISBN 978-0-7844-1540-5 (print)
ISBN 978-0-7844-8240-7 (PDF)
Manufactured in the United States of America.

25 24 23 22 21 20 19 1 2 3 4 5

ASCE STANDARDS

The Board of Direction approved revisions to the ASCE Rules for Standards Committees to govern the writing and maintenance of standards developed by ASCE. All such standards are developed by a consensus-standards process managed by the ASCE Codes and Standards Committee (CSC). The consensus process includes balloting by a balanced standards committee and reviewing during a public comment period. All standards are updated or reaffirmed by the same process at intervals between five to ten years. Requests for formal interpretations shall be processed in accordance with Section 7 of ASCE Rules for Standards Committees, which are available at http://www.asce.org/uploadedFiles/Technical_Areas/Codes_and_Standards/Content_Pieces/asce-rules-standards-committees.pdf.

Errata, addenda, supplements, and interpretations, if any, for this standard can also be found at http://ascelibrary.org/.

The provisions of this document are written in mandatory language and, as such, are intended to be suitable for regulatory or contractual purposes.

This standard has been prepared in accordance with recognized engineering principles and should not be used without the user's competent knowledge for a given application. The publication of this standard by ASCE is not intended to warrant that the information contained herein is suitable for any general or specific use, and ASCE takes no position respecting the validity of patent rights. The user is advised that the determination of patent rights or risk of infringement is entirely his or her own responsibility.

DEDICATIONS

Robert P. Kennedy, Ph.D., P.E., NAE
April 2, 1939–December 30, 2018

ASCE 43-19 is dedicated to Dr. Robert (Bob) Phillip Kennedy, a leader in the global nuclear energy industry for more than 40 years, with seminal contributions in civil and structural engineering, and probabilistic risk assessment.

Bob enrolled at Stanford University in 1957 where he graduated 10 years later with three degrees: Bachelor of Science (1960), Master of Science (1961), and Doctor of Philosophy (1967). His undergraduate studies were supported in part by a Holmes and Narver scholarship and he worked summers for this consultancy in Los Angeles. A keen athlete, he was on the swimming and water polo teams through graduate school. He graduated Phi Beta Kappa. He continued his studies at Stanford on a graduate fellowship from the US Department of Defense under the direction of Professor Jack Benjamin. While a graduate student, Bob cotaught a class on blast-resistant design with his Stanford classmate and fellow Benjamin advisee, the late C. Allin Cornell—the start of a lifelong collaboration between two of the most productive, respected engineers of their generation.

In May 1964, before finishing his Ph.D., Bob joined the US Army Corps of Engineers. He went to Ft. Belvoir in Virginia, where, after basic training, he was placed in charge of engineering courses for officers-in-training. In August 1965, he volunteered to move to Okinawa, Japan, to work on the design and construction of the Cam Rahn Bay logistics facility for the US Navy, an experience he described as unique and invaluable but one that gave him no time to complete his dissertation. Bob joined Holmes and Narver in Los Angeles in May 1966, after being discharged from the Corps, working there for two years. He filed his Ph.D. dissertation, "A statistical analysis of the shear strength of reinforced concrete beams" in April 1967. He relocated to the Holmes and Narver office in Las Vegas for two years to support work related to underground nuclear detonations at the Nevada Test Site. He left Holmes and Narver in 1977 to start the Southern California office of the Engineering Decision and Analysis Company (EDAC). He left EDAC in 1980 and established Structural Mechanics Associates (SMA) with a group of like-minded engineers. Bob preferred technical work to managing a multi-office engineering consultancy and SMA was sold in 1984. He provided independent engineering consultation at RPK Structural Mechanics Consulting, Inc. for the following 34 years.

As an engineering consultant, Bob worked in many subject areas, mostly involving dynamic analysis and design and construction of mission-critical structures for the nuclear, defense, industrial, and petroleum industries. His work on the seismic ruggedness of nuclear facilities and the quantification of earthquake-related risk was fundamental, included the development of performance-based seismic design criteria for DOE Standard 1020 and ASCE 43; fragilities of structures, systems and components for use in seismic probabilistic risk and seismic margins assessments; methodologies for seismic walkdown reviews of critical facilities, including nuclear power plants; and teaching short courses on seismic probabilistic risk assessment around the world. He chaired the Senior Seismic Review and Advisory Panel from 1983 to 1992, providing guidance to the USNRC and power utilities on the seismic robustness of equipment in existing nuclear power plants. He served as lead author of a number of seminal reports in the 1980s and 1990s, which are still widely used today, related to seismic risk and margin assessments, including EPRI NP-6041 (August 1991), EPRI TR-103959 (April 1994), NUREG/CR-4334 (August 1985) and NUREG/CR-5270 (March 1989). Less well known but equally important technical contributions by Bob were in the domain of hardened structures and equipment, through consulting to the US Defense Threat Reduction Agency, US Defense Nuclear Agency, and other US government entities on a broad range of subjects, mostly related to extreme ground shock, pressures and temperatures generated by underground nuclear detonations.

Bob was an active member of the ASCE Committee on Dynamic Analysis of Nuclear Structures and its predecessor committee for nearly 40 years. He served as the first chair of the committee, which produced ASCE Standard 4-86. His pioneering work on performance-based seismic design criteria for safety-related nuclear facilities in the 1990s underpinned the technical basis for ASCE 43-05, enabling deterministic design procedures to be used to meet probabilistic goals. He made numerous other contributions to this standard and its analysis counterpart, ASCE 4, including characterization of ground shaking, modeling, analysis methodologies, soil–structure interaction, and seismic isolation. Bob improved this 2019 edition of ASCE 43 by revising the definition of earthquake input, replacing the design factor he introduced in ASCE 43-05 with a more broadly applicable scale factor. He also contributed to the writing of other codes and standards, including ACI 349, with an emphasis on design procedures for impactive and impulsive loadings on nuclear structures, and ASME standards for power piping.

Bob received many awards and honors over his career. He was elected to the US National Academy of Engineering in 1991 for "developing design procedures for civil and mechanical structures to resist seismic and other extreme loading conditions." Other awards included the 1992 ASCE Stephen Bechtel Energy Engineering Award and the 2015 ASCE LeVal Lund Award. Bob was a Fellow of the California Council of Science and Technology.

ASCE and the members of its ASCE DANS committee in particular, the global civil and structural engineering community, and all tasked with the design of energy-related facilities owe a debt of thanks to Bob Kennedy that cannot be repaid. Instead, and as a token of our deep appreciation and thanks, we dedicate this 2019 edition of ASCE 43 to our friend and colleague. Bob's absence from DANS committee discussions and deliberations is, and will be, sadly missed.

Stephen A. Short, P.E.
February 12, 1945–October 14, 2016
"It is nice to be important, but it is more important to be nice."

ASCE 43-19 is dedicated to Mr. Stephen Alan Short, a leader in the development of codes and standards for ASCE, a dedicated structural engineering professional, a loving husband and father, and a friend to many. The above quote from John Templeton circa 1912 was one of Steve's favorites and exemplified how he lived his life.

Steve received his Bachelor of Science and Master of Science degrees from the University of California, Berkeley, in 1966 and 1968, respectively. His studies focused on the dynamic analysis of structures and structural design. He began his professional career at Teledyne in 1968. He left Teledyne and joined Holmes and Narver in Anaheim, California, where he specialized in the design and analysis of line of sight vent line closures for the support of the underground nuclear program at the Nevada Test Site. In 1974, he joined to the Engineering Decision and Analysis Company (EDAC) in Southern California. In 1983, he joined Structural Mechanics Associates (SMA) where his work focused on dynamic analysis of structures for the effects of earthquake shaking and blast loadings. He left (SMA) in 1991 to form a Southern California office of ABB/Impell, Inc. In 1995, he left Impell to join EQE, Inc. where he formed close friendships with a number of colleagues with like interests. He remained with EQE until joining Simpson Gumpertz & Heger in 2008. During his career, Steve worked on soil–structure interaction analysis of large civil structures, dynamic analysis of structures and components, and evaluation of air-blast effects, and also developed codes and also standards such as ASCE 4 and ASCE 43.

Steve was an active member of the ASCE Committee on Dynamic Analysis of Nuclear Structures and its predecessor committee for nearly 40 years. He worked with Dr. Bob Kennedy to develop the probabilistic basis for much of the work adapted by the US Department of Energy in UCRL-15910, Design and Evaluation Guidelines for Department of Energy Facilities Subjected to Natural Phenomena Hazards, which later formed the foundations of ASCE 4-16 and ASCE 43-05. Steve was instrumental in promulgating these standards to the engineering community through DOE workshops and site-specific training.

Steve's influence reaches far beyond his contributions to civil engineering. He displayed a humble spirit, a tremendous work ethic, and a willingness to help others. Steve had a tremendously infectious, positive personality. He was very quick to smile or to make a joke. He was easy to talk with and extremely patient. As a senior member of the DANS committee and as a mentor to many, Steve never neglected to take the time to sit with younger engineers to work solutions to difficult problems. He never lost sight of the fact that civil engineering solutions ultimately serve to improve our quality of life.

Steve loved his family first, then sports. He was married to his wife, Gale, for 43 years and had three children. He never missed his children's many games, including football, baseball, soccer, cheerleading, or swim meets. He was an extremely supportive husband and father and was always where he needed to be. He played basketball, volleyball, and was a downhill skier and a golfer. Steve was an avid fan of the California Golden Bears, the San Francisco 49ers, and the San Diego Chargers. He could recite statistics about baseball and football that amazed all, as that was one of his passions. He constantly cheered for the underdog, which may explain why he loved Cal football so much. He also loved music, and during his lifetime he attended hundreds of musicals and concerts. But his professional life was also extremely important to him and during his last few years of life under treatment, when he could have stayed home, he continued to do his job, never missing a day on which he could work in 46 years.

ASCE and the members of its ASCE DANS committee dedicate this 2019 edition of ASCE 43 to our friend and colleague, Stephen A. Short. Steve's contributions to ASCE 4 and 43 have been, are, and will be, sorely missed.

David K. Nakaki, Ph.D., P.E.
August 3, 1957 – December 9, 2019

ASCE 43-19 is dedicated to Dr. David Kiyoshi Nakaki, an expert in seismic/structural engineering for nuclear facilities, a devoted father, and a friend to many.

Dave was born and raised in West Los Angeles. He received all of his college degrees at nearby University of California at Los Angeles; Bachelor's of Science in civil engineering in 1980, Master's of Science in engineering in 1983, and Doctor of Philosophy in engineering in 1987. His doctoral thesis was titled "Uplifting Response of Structures Subjected to Earthquake Motions."

Upon his university graduation, Dave began his professional career in 1987. His career growth was represented by steady rise consistent with his talents through employment at NTS Engineering (1987 – 1988), Impell Corporation (1988 – 1991), EQE International / ABS Consulting (1991 – 2003), ARES Corporation (2003 – 2008), and Simpson Gumpertz & Heger Inc. (SGH) (2008 – 2019). Dave was a key member of the core group that founded the Orange County, California, office of SGH. He was a licensed civil engineer in the states of California and Alaska.

Dave's professional career was dedicated to ensuring public safety through the seismic evaluation of commercial nuclear power plants and US Department of Energy (DOE) facilities for design basis and beyond design basis earthquake events. He was an expert in all technical elements of such evaluations, including soil-structure interaction modeling and analysis, seismic response analysis, by deterministic and probabilistic methods, seismic capacity evaluation of structures, and experience-based seismic walkdown assessments of essential equipment. Dave was a longtime member of the ASCE Dynamic Analysis of Nuclear Structures Committee. He was the principal contributor to the provisions for probabilistic site response analysis, which were first included in ASCE/SEI 4-16.

Dave was responsible for beyond design basis evaluations of nuclear power plants by seismic probabilistic risk assessment (SPRA) and seismic margin assessment (SMA). His greatest achievement in this arena was completion of the Diablo Canyon Power Plant SPRA, a challenging project because of the high site seismicity. Dave also led SPRAs and SMAs in several other countries, including Canada, France, Germany, and Switzerland. He was responsible for seismic evaluations of DOE nuclear facilities implementing ASCE 4 and ASCE 43 at the Idaho National Laboratory, Los Alamos National Laboratory, Nevada Nuclear Security Site, Oak Ridge National Laboratory, and Pantex Plant.

Over the last several years of his career, Dave was essentially the Renaissance man of his engineering group. Besides providing project management leadership and being the "go-to" person for guidance on technical subjects, he was responsible for supervising younger engineers, recruiting new engineering talent, computer hardware and software management, and quality assurance. Dave seldom declined to provide assistance to others when asked. In those rare occasions when he did, it was only because somebody else convinced him he already had too much to do. Dave was particularly an excellent mentor of young engineers. He was a patient teacher always willing to offer advice and answer questions when asked. He made great contributions to development of the future leadership in our industry.

In the workplace, Dave was a treasured friend and colleague to all. He served as a great role model for the next generation of engineers through his strong work ethic. While he could present a private and reserved image, there were occasions when he would break out his incredible wit to the enjoyment of those around him. Conversations with Dave would also display his depth of knowledge of the world around him that went well beyond engineering. His presence is missed every day.

Dave was a dedicated family man. A devoted father to three children, Karen, Mitchell, and Claire, he spent many years coaching youth basketball or timing, stopwatch in hand at swim meets. More recently, he also stepped up and spent time each week with his mother. Dave was a proud Japanese American and could make the mochi to prove it. As he did with everything in his life, Dave took the time to perfect his Thanksgiving turkey and Christmas roast recipes and could find the best restaurants – sometimes hole-in-the-wall burgers, other times old school Chinese food or the freshest sushi. Perhaps most indicative of Dave's personality was his loyalty to the Grateful Dead. Beginning in his college years, Dave attended numerous live shows and wound up with hundreds of live recordings on vinyl, cassette tapes, and CDs.

CONTENTS

COMMENTARY TO ASCE 43-19

PREFACE

Nuclear facilities process, store, or handle radioactive materials in a form and quantity that pose potential hazard to the workers, the public, or the environment. Because of the risk associated with such hazards, it is desirable that nuclear facilities have a lower probability of structural damage caused by earthquake shaking than commercial buildings. The goal of this standard is to ensure that nuclear facilities are properly designed to withstand the effects of earthquake ground shaking while retaining the desired functionality, expressed as target performance goals. Design for other earthquake effects (such as differential fault displacement and seismic slope instability) is not covered by this standard. This standard is intended for use in the design of new facilities and should be used in conjunction with other national consensus standards specified herein. This standard can be used for assessment of existing facilities.

This standard can also be used for facilities handling explosives, toxic materials, or chemicals, and for facilities where safety, mission, or investment protection is an explicit design goal.

This standard is intended to be used with ASCE/SEI Standard 4, which provides criteria for seismic analysis of safety-related nuclear structures, systems, and components (SSCs); ACI standards for concrete structures; AISC standards for steel structures ANS Standards for design, analysis, and operation of components, systems, and facilities utilizing nuclear technology; ASME standards for mechanical systems and components; and IEEE standards for electrical systems and components. The writers of this standard sought to integrate its provisions and commentary with that of ASCE/SEI Standard 4-16.

This standard includes a new chapter on the design of seismically isolated nuclear facilities, and includes provisions for prototype and production testing of isolators. Provisions and commentary are adapted in part from Chapter 12 of ASCE/SEI Standard 4-16.

The intended user of this standard is the engineer or analyst involved in the design of new, or the assessment of existing, nuclear structures, systems, or components.

ACKNOWLEDGMENTS

ASCE acknowledges the work of the Nuclear Standards Committee and the Task Committee on Dynamic Analysis of Nuclear Structures, both under the auspices of the Codes and Standards Activities Division of the Structural Engineering Institute.

Nuclear Standards Committee
Andrew S. Whittaker, Ph.D., P.E., S.E., F.ASCE, F.SEI, *Chair*
F George Abatt, Ph.D., P.E., F.ASCE, *Vice Chair*
Lisa M. Anderson, P.E., F.ASCE, F.SEI
Michael Cohen, P.E., M.ASCE
Justin L. Coleman, Ph.D., P.E., M.ASCE
Benjamin Kosbab, Ph.D., P.E., M.ASCE
Brian M. McDonald, Ph.D., P.E., F.ASCE
Gregory E. Mertz, Ph.D., P.E., F.ASCE
Farhang Ostadan, Ph.D., P.E., F.ASCE
Josh Parker, P.E., M.ASCE
Michael W. Salmon, P.E., F.ASCE
Neda Stoeva, P.E., M.ASCE
Mohamed M. Talaat, Ph.D., P.E., S.E., M.ASCE
Derrick A. Watkins, Ph.D., P.E., M.ASCE
Jim Xu, Ph.D.

Task Committee on Dynamic Analysis of Nuclear Structures
Voting Members
Michael W. Salmon, P.E., F.ASCE, *Chair*
F George Abatt, Ph.D., P.E., F.ASCE, *Vice Chair*
Neda Stoeva, P.E., M.ASCE, *Secretary*
Jeremy Aartun Sr., P.E., S.E.
Lisa M. Anderson, P.E., F.ASCE, F.SEI
Said Bolourchi, Ph.D., F.ASCE
Shawn A. Carey, Ph.D., P.E., M.ASCE
Justin L. Coleman, Ph.D., P.E., M.ASCE
Carl J. Costantino, Ph.D., P.E., F.ASCE
Michael C. Costantino, P.E., M.ASCE
Daniel W. Eggers, P.E., M.ASCE
Dan M. Ghiocel, Ph.D.
Philip S. Hashimoto, P.E., F.ASCE
Quazi A. Hossain, Ph.D., P.E., F.ASCE
Thomas W. Houston, Ph.D., P.E., F.ASCE
Roy J. Hunt, P.E., M.ASCE
Stuart Jensen
James J. Johnson, Ph.D., P.E., F.ASCE
Robert P. Kennedy, Ph.D., P.E., M.ASCE (deceased)
Mohsin R. Khan, Ph.D., P.E., M.ASCE
Gunup Kwon, Ph.D., P.E., M.ASCE
Eric R. Macfarlane, P.E., S.E., M.ASCE
Sanjeev R. Malushte, Ph.D., P.E., S.E., F.ASCE
Brian M. McDonald, Ph.D., P.E., S.E., F.ASCE
Greg E. Mertz, Ph.D., P.E., F.ASCE
David K. Nakaki, Ph.D., P.E., M.ASCE (deceased)
Farhang Ostadan, Ph.D., P.E., F.ASCE
Raj S. Rajagopal, D.Eng., P.E., F.ASCE
Jaspal S. Saini, P.E., M.ASCE
Subir K. Sen, P.E., M.ASCE
Stephen A. Short, P.E., M.ASCE (deceased)
Mohamed M. Talaat, Ph.D., P.E., S.E., M.ASCE
Raman M. Venkata, P.E., F.ASCE
Shen Wang, Ph.D., P.E., M.ASCE
Derrick A. Watkins, Ph.D., P.E., S.E., M.ASCE
William White, Ph.D., P.E., M.ASCE
Andrew S. Whittaker, Ph.D., P.E., S.E., F.ASCE, F.SEI
Jim Xu, Ph.D.

Associate Members
Abhinav Anup, P.E., M.ASCE
Alejandro P. Asfura, Ph.D.
Chandrakanth Bolisetti, Ph.D., A.M.ASCE
Nicholas Brown, Ph.D., A.M.ASCE
Parthasarathy Chandran
Michael Cohen, P.E., M.ASCE
Isabel Cuesta, Ph.D., P.E., M.ASCE
Amitabh Dar, P. Eng.
Samer M. El-Bahey, Ph.D., P.E., M.ASCE
Eddie M. Guerra, P.E., M.ASCE
Ajaya K. Gupta, Ph.D., F.ASCE
Alidad Hashemi, Ph.D., P.E., S.E., M.ASCE
Yigit Isbiliroglu, Ph.D., P.E., M.ASCE
Boris Jeremic, Ph.D.
Eric S. Keldrauk, D.Eng., P.E., M.ASCE
Minkyu Kim, Ph.D.
Ben Kosbab, Ph.D., P.E., M.ASCE
Andrew S. Maham, P.E., M.ASCE
Troy A. Morgan, Ph.D., P.E., M.ASCE
Mustafa Ozkan, P.E., M.ASCE
Si-Hwan Park, Ph.D., M.ASCE
Josh Parker, P.E., M.ASCE
Ayman Saudy, Ph.D., P.Eng, M.ASCE
Robert E. Spears Ph.D.
Luben I. Todorovski, Ph.D., P.E., M.ASCE
Jenna Wong, Ph.D., P.E., M.ASCE
Zia Zafir, Ph.D., G.E., P.E., M.ASCE

CHAPTER 1
INTRODUCTION

1.1 ACRONYMS AND ABBREVIATIONS

ACI	American Concrete Institute
AISC	American Institute of Steel Construction
AISI	American Iron and Steel Institute
ANS	American Nuclear Society
ANSI	American National Standards Institute
APE	Annual probability of exceedance
ASCE/SEI	American Society of Civil Engineers/Structural Engineering Institute
ASD	Allowable stress design
ASME	American Society of Mechanical Engineers
ATC	Applied Technology Council
AWWA	American Water Works Association
B&PVC	Boiler and Pressure Vessel Code
BDBE	Beyond design basis earthquake
BRBF	Buckling restrained braced frame
CBF	Concentrically braced frame
CFR	Code of Federal Regulations
CMAA	Crane Manufacturers Association of America
COV	Coefficient of variation
CS	Clearance to the stop
DBE	Design basis earthquake
DF	Design factor
DOE	US Department of Energy
DRS	Design response spectrum
EBF	Eccentrically braced frame
EES	Earthquake experience spectrum
ENA	Eastern North America
EOR	Engineer of record
EPRI	Electric Power Research Institute
EUS	Eastern United States
FEMA	Federal Emergency Management Agency
FP	Friction pendulum
FS	Factor of safety
GERS	Generic equipment ruggedness spectrum
GIP	Generic implementation procedure
HVAC	Heating, venting, and air conditioning
IBC	International Building Code
IEEE	Institute of Electrical and Electronics Engineers
ISRS	In-structure response spectrum
LDR	Low damping rubber
LR	Lead rubber
LRFD	Load and resistance factor design
LS	Limit state
M/D	Magnitude and distance pair
NEHRP	National Earthquake Hazards Reduction Program
NEMA	National Electrical Manufacturers Association
NEP	Nonexceedance probability
NFPA	National Fire Protection Association
NPH	Natural phenomena hazards
NPP	Nuclear power plant
NRC	Nuclear Regulatory Commission
NUREG	US Nuclear Regulatory Commission technical report designation
NUREG/CR	Contractor-prepared NUREG
OCBF	Ordinary concentrically braced frame
PC	Performance category
PGA	Peak ground acceleration
PSD	Power spectral density
PSHA	Probabilistic seismic hazard assessment
QA	Quality assurance
QME-1	Qualification of active mechanical equipment used in nuclear facilities
RBS	Reduced beam section
RRS	Required response spectrum
SA	Spectral acceleration
SAM	Seismic anchor motion
SC	Steel-plate composite
SCBF	Special concentrically braced frame
SDB	Seismic design basis
SDC	Seismic design category
SF	Scale factor
SMACNA	Sheet Metal and Air Conditioning Contractors' National Association
SMRF	Special moment-resisting frame
SQUG	Seismic Qualification Utility Group
SQURTS	Seismic qualification reporting and testing standardization
SRSS	Square root of the sum of squares
SSC	Structure, system, and component
SSE	Safe shutdown earthquake
SSI	Soil–structure interaction
TES	Test experience spectrum
TRS	Test response spectrum
UHRS	Uniform hazard response spectrum
USGS	US Geological Survey
V&V	Verification and validation
ZPA	Zero-period acceleration

1.2 DEFINITIONS

Active Component: A component that must change state as part of its safety function during an earthquake.

Arias Intensity: A measure of intensity of ground shaking calculated by integrating the square of the ground acceleration over a specified period of time.

Arias Intensity Rise Time: Time to generate 5% of the total cumulative energy in an earthquake time series.

Basemat: In the context of seismically isolated structures, the basemat is a thick, reinforced concrete diaphragm immediately above the isolation system.

Bearing Wall System: A structural system without a complete gravity load-carrying space frame. Resistance to horizontal earthquake shaking is provided by shear walls or braced frames.

Building Frame System: A structural system with a complete gravity load-carrying space frame. Resistance to horizontal earthquake shaking is provided by shear walls or braced frames.

Clearance to the Stop: The horizontal distance between the superstructure of a seismically isolated structure and the stop.

Design Basis Earthquake (DBE): The description of the ground motion, defined in terms of the DRS, to be used for design.

Design Response Spectrum (DRS): The product of a UHRS and a scale factor.

Design Team: The responsible group charged with producing the design.

Directional Correlation Coefficient: The cross-correlation coefficient for two orthogonal time series.

Distribution System: A system (i.e., collection of components) whose function is to distribute material or data (e.g., fluid, signals, or power). Examples are piping, cable trays, conduit, and HVAC systems.

Earthquake Experience Spectrum: The capacity of a reference equipment class described using a response spectrum, based on earthquake experience data.

Effective Natural Frequency: The frequency of the mode that dominates response.

Effective Seismic Weight: The dead load and other permanent loads, such as stationary equipment, piping, and fractions of the live load and snow load, as described in Section 3.4.2 of ASCE 4-16 (2017b).

Equivalent Horizontal Stiffness (Secant Stiffness): In the context of a seismically isolated structure, the value of the lateral force in an isolation system, or an element thereof, divided by the corresponding lateral displacement; also termed *Secant stiffness*.

Equivalent Viscous Damping Ratio: In the context of a seismically isolated structure, the value of the equivalent viscous damping corresponding to energy dissipated during cyclic response of a seismic isolation system.

Foundation: In the context of a seismically isolated structure, a foundation is the reinforced concrete construction, including pedestals, below the isolators that support the isolators.

Generic Equipment Ruggedness Spectrum (GERS): A generic spectrum that represents the capacity of a reference equipment class as demonstrated by test. The GERS is intended to represent the highest input level for which one has high confidence of successful equipment performance.

Graded Approach: A process by which the level of analysis, documentation, and actions necessary to comply with requirements are commensurate with

- Relative importance to safety, safeguards, and security, and of radiological and nonradiological hazards;
- Magnitude of any hazard involved;
- Life cycle stage of a facility;
- Programmatic mission of a facility;
- Particular characteristics of a facility; and
- Other relevant factors.

Hazard: An event or natural phenomenon that poses some risk to a facility. Internal hazards include events such as equipment failures, human failures, and flooding and fires internal to the plant. External hazards include events such as flooding and fires external to the plant, tornadoes, earthquakes, and aircraft crashes.

In-structure Response Spectra (ISRS): The response spectra generated from the dynamic response of the structure at selected locations in a structure. In-structure response spectra are used for design of systems and components supported within a structure.

Inelastic Energy Absorption Factor: A factor used to reduce seismic demand to account for inelastic behavior. The inelastic energy absorption factor is a function of the limit state and the structural system or equipment configuration.

Isolation Gap (Moat): In the context of a seismically isolated structure, the moat or isolation gap is the width around the perimeter of the isolated superstructure in which the superstructure can move without restriction. The width is defined by the clearance to the stop.

Isolation Interface: In the context of seismically isolated structures, the isolation interface is the interface between the isolated superstructure and the supporting (nonisolated) foundation.

Isolation System: In the context of a seismically isolated structure, the isolation system is the collection of structural elements that includes all individual isolator units, all structural elements that transfer force between elements of the isolation system, and all connections to other structural elements.

Isolation System Effective Damping: In the context of seismic isolation systems, the isolation system effective damping is the equivalent viscous damping based on isolator hysteretic damping, and it corresponds to the energy dissipated during cyclic response of the isolation system.

Isolator Unit: In the context of seismic isolation systems, an isolator unit is a horizontally flexible and vertically stiff structural element of the isolation system that permits large lateral deformations under design seismic load. An isolator unit may be used either as part of, or in addition to, the weight-supporting system of the structure.

Limit State (LS): The limiting acceptable condition of the structure, system, or component. The limit state may be defined in terms of a maximum acceptable displacement, strain, ductility, or stress. Four limit states are defined for nuclear safety-related SSCs.

Mean Annual Hazard Exceedance Frequency: The expected annual frequency of exceeding a ground motion spectral acceleration at a given site for a user-specified spectral frequency.

Moat (Isolation Gap): In the context of a seismically isolated structure, the moat or isolation gap is the width around the perimeter of the isolated superstructure in which the superstructure can move without restriction. The width is defined by the clearance to the stop.

Moment-Resisting Frame System: A structural system with a complete gravity load-resisting space frame. Resistance to horizontal earthquake shaking is provided by axial-flexural action of its beams and columns.

Nonbuilding Structure: Nonbuilding structures include all self-supporting structures other than buildings. Nonbuilding structures shall be designed to provide the strength and stiffness required to resist the seismic forces specified in this standard.

Nyquist Frequency: The maximum frequency that can be represented by a time series.

Passive Component: A component that does not have to change state as part of its safety function during an earthquake.

Peak Ground Acceleration (PGA): The maximum absolute value of a component of a ground acceleration time series triplet.

Peer Review: A formal review process in which an external party reviews the methodology, results, and process by which a design is developed or an evaluation is carried out. The external party shall be independent of project schedule and budget constraints.

Power Spectral Density (PSD): A distribution of power in a time series, as a function of frequency.

Probabilistic Seismic Hazard Assessment (PSHA): A procedure used to develop seismic hazard curves at a site.

Required Response Spectrum (RRS): The response spectrum that is used to qualify an SSC.

Scale Factor: A factor, less than 1.0, that multiplies the ordinates of the UHRS calculated for the annual frequency of exceedance of the target performance goal to determine the DRS.

Secant Stiffness (Equivalent Horizontal Stiffness): In the context of a seismically isolated structure, the value of the lateral force in an isolation system, or an element thereof, divided by the corresponding lateral displacement; also termed *equivalent horizontal stiffness*.

Seismic Anchor Motion: Seismically induced relative motion between distinct anchor points of an SSC.

Seismic Design Basis (SDB): The combination of seismic design category (1, 2, 3, 4, or 5) and limit state (A, B, C, or D) that determines the design basis earthquake and acceptance criteria for designing SSCs. For example, Seismic Design Basis 3C uses criteria given in this standard for Seismic Design Category 3 and Limit State C.

Seismic Design Category (SDC): A category assigned to an SSC that is a function of the severity of adverse radiological and toxicological effects on workers, the public, and the environment that may result from the seismic failure of the SSC. SSCs may be assigned to seismic design categories that range from 1 to 5. For example, a conventional building whose failure may not result in any radiological or toxicological consequences is assigned to Seismic Design Category 1; a safety-related SSC in a facility where nuclear material is processed with a large inventory of radioactive material may be placed in Seismic Design Category 5. In this standard, the term *Seismic Design Category* has a different meaning from that in the International Building Code.

Seismic Hazard Curve: A product of PSHA that plots, for a given spectral frequency, spectral acceleration versus mean annual frequency of exceedance.

Slab-wall Moment Frame: A moment-resisting frame composed of walls and slabs.

SQUG Reference Spectrum: A generic earthquake experience spectrum developed during the Seismic Qualification Utility Group program that represents a level of shaking at which a reference equipment class is expected to perform successfully.

Stop: A stop is a structure or series of structures designed to prevent excessive displacement of the isolation system. A moat wall could serve as a stop.

Structure, System, and Component (SSC): A *structure* is an element, or a collection of elements, to provide support or enclosure, such as a building, freestanding tanks, basins, dikes, or stacks. A *system* is a collection of components assembled to perform a function, such as piping, cable trays, conduits, or HVAC. A *component* is an item of mechanical or electrical equipment, such as a pump, valve, or relay, or an element of a larger array, such as a length of pipe, elbow, or reducer.

Superstructure: In the context of a seismically isolated structure, the superstructure is composed of all structural elements above the isolation system (e.g., slabs, beams, columns, and walls). For a conventional light-water reactor, the structural framing includes primary and secondary containment, internal structure to support the power generation and safety-related components and systems, and the basemat (or diaphragm) immediately above the isolation system.

Target Performance Goal: Target annual frequency of exceeding a specified limit state.

Test Experience Spectrum: Test-based response spectrum that defines the seismic capacity of a reference equipment class.

Test Response Spectrum: Response spectrum that corresponds to the actual motion of a shake table.

Umbilical Lines: In the context of a seismically isolated structure, umbilical lines are nonstructural components and systems, mainly distribution systems, which cross the isolation interface and sustain the large isolator displacements (or deformations) associated with design basis and beyond design basis earthquake shaking. Examples of umbilical lines could include high-pressure steam lines from the power reactor to the turbines and cables located on trays or in ducts from emergency power systems located off the nuclear island to the power reactor. In the context of rocking and sliding components, umbilical lines are distribution systems that bridge between the component and the supporting structure.

Uniform Hazard Response Spectra (UHRS): Response spectra derived so that the annual frequency of exceeding the spectral quantity (e.g., acceleration or displacement) is the same for all spectral frequencies.

Validation: The process of determining the degree to which a verified model is an accurate representation of the real world (e.g., phenomenon, system, process, or problem of interest) from the perspective of the intended uses of the model.

Verification: The process of determining that a computational model accurately represents the underlying mathematical model and its solution.

1.3 SEISMIC DESIGN CRITERIA

This standard, when integrated with ASCE 1, ASCE 4, and other referenced industry design, detailing, and construction standards, provides criteria for seismic design of safety-related structures, systems, and components (SSCs) to achieve target performance goals in nuclear facilities. The intended users of this standard are civil, mechanical, and structural engineers designing or evaluating safety-related nuclear facilities.

Graded seismic design criteria are presented in this standard. More stringent design criteria are used to achieve better seismic performance for SSCs that have more serious failure consequences, as recognized by the designation of a seismic design basis (SDB), which varies by SSC. The SDB for nuclear facilities is fully defined by (1) a qualitative description of the acceptable level of damage, denoted here as a limit state (LS), and (2) a seismic design category (SDC), where SDC and LS are component-specific and are assigned according to ANS 2.26 (2017). Table 1-1 summarizes the design provisions for the four SDCs that are the subject of this standard: 2, 3, 4, and 5. Selection of the appropriate SDC and LS for a given structure or usage is beyond the scope of this standard, which provides design requirements for a given SDB, once selected.

The annual frequency of exceeding the acceptable performance level is called the *target performance goal*, and it decreases as SDC increases. This decrease is achieved by increasing the return period of the design basis earthquake (DBE) shaking. The DBE ground motion is determined by amplitude scaling uniform hazard response spectra (UHRS), as seen in Table 1-1. Underpinning the procedures described in this standard are two key assumptions, namely, (1) the analysis procedures of ASCE 4 predict seismic demands for DBE shaking at the 80th percentile nonexceedance probability, and (2) the equations for design strength in material standards ACI 349 (2013b) and AISC N690 (2018) deliver capacities at the 98th percentile exceedance probability. Used together, the intent is to achieve a 1% or smaller annual frequency of unacceptable performance

Table 1-1. Summary of Earthquake Design Provisions.

	Seismic Design Category			
	2	3	4	5
Target performance goal, P_F	4×10^{-4}	1×10^{-4}	4×10^{-5}	1×10^{-5}
DBE response spectrum or acceleration time series	SF × UHRS; Chapter 2 in this standard			
Damping for structural evaluation	Section 3.3.3			
Analysis methods for structures	ASCE 4 and Chapter 3 in this standard			
Analysis methods for systems and components	In-structure response spectra; ASCE 4 and Chapter 8 in this standard			
Load factor	1.0			
Inelastic energy absorption factors	Table 5-1 and/or Table 8-1 in this standard			
Material strength	Minimum specified value			
Component design strength	Design strength according to materials standards unless exceptions are made in this standard			
QA program	Chapter 10 in this standard			
Independent peer review	Chapter 10 in this standard			

Table 1-2. Deformation and Damage by Limit State.

Limit State	Expected Deformation	Expected Damage
A	Large permanent distortion, short of collapse	Significant
B	Moderate permanent distortion	Generally repairable
C	Limited permanent distortion	Minimal
D	Essentially elastic behavior	Negligible

Source: Adapted from ANS 2.26 (ANS 2017).

conditioned on DBE shaking, and a 10% or smaller annual frequency of unacceptable performance conditioned on 150% of DBE shaking.

For a given DBE shaking, the damage is greatest for Limit State A and negligible for Limit State D (Table 1-2). Damage is limited to allowable levels by applying inelastic energy absorption factors and/or deformation limits, as specified in Chapters 5 and 8. Design criteria are presented in Chapters 2 through 9. Requirements for quality assurance (QA) and independent peer review are described in Chapter 10.

1.4 INTEGRATION OF OTHER CODES AND STANDARDS WITH ASCE 43

Consensus codes and standards, which follow procedures approved by ANSI and support the design process, are published by a number of professional organizations including the American Concrete Institute (ACI), American Institute of Steel Construction (AISC), American Nuclear Society (ANS), ASCE, American Society of Mechanical Engineers (ASME), and Institute of Electrical and Electronics Engineers (IEEE).

ASCE publishes three standards related to the seismic design of nuclear facilities: ASCE 1, *Guideline for Design and Analysis of Nuclear Safety-Related Earth Structures*; ASCE 4, *Seismic Analysis of Safety-Related Nuclear Structures*; and this standard, ASCE 43.

The American Concrete Institute publishes ACI 349, *Code Requirements for Nuclear Safety-Related Concrete Structures*, to be used for the design and detailing of reinforced concrete nuclear structures. The American Institute of Steel Construction publishes AISC N690, *Specification for Safety-Related Steel Structures for Nuclear Facilities*. Some exceptions and additions to ACI 349 and AISC N690 are required by this standard.

The American Nuclear Society Standards ANS 2.27 and ANS 2.29 describe the development of seismic hazard curves, which define ground motion parameters as a function of return periods or exceedance frequencies. ANS 2.26 establishes five SDCs, introduced earlier in this chapter, which are defined using consequences of failure as measured by unmitigated radiological exposure to facility workers and the public. ANS 2.26 describes four limit states for SSCs in all five SDCs.

The American Society of Mechanical Engineers and IEEE publish standards that use ASCE-calculated seismic loads for design and provide procedures to determine the effects of these loads on SDC 1 through SDC 5 systems, components, and acceptance criteria. Standards that are used for the seismic design of mechanical and electrical systems and components include those of the American Petroleum Institute (API), Sheet Metal and Air Conditioning Contractors' National Association (SMACNA), American Water Works Association (AWWA), Crane Manufacturers Association of America (CMAA), and National Electrical Manufacturing Association (NEMA).

1.5 ALTERNATIVE METHODS TO MEET THE INTENT OF THIS STANDARD

The intent of this standard is to achieve the performance goals of Table 1-1 for the limit states defined in Table 1-2. Alternative methods may be used to achieve a performance goal, but they must justify (and properly incorporate into reliability calculations) appropriate site-specific hazard curves, demands, and capacities, with explicit consideration of uncertainty and variability. One method to comply is to reasonably achieve both of the following: (1) less than a 1% probability of unacceptable performance for the DBE shaking defined in Chapter 2, and (2) less than a 10% probability of unacceptable performance for 150% DBE shaking. Acceptable alternate methods for achieving a target performance goal, given capacities defined at different exceedance levels, are presented in the nonmandatory Appendixes A and B. Designs or evaluations based on alternative methods allowed in this section shall be independently peer-reviewed.

CHAPTER 2
EARTHQUAKE GROUND MOTION

2.1 SEISMIC HAZARD EVALUATION

The design basis earthquake (DBE) shall be based on seismic hazard curves and uniform hazard response spectra (UHRS) from a probabilistic seismic hazard assessment (PSHA). The hazard curves and UHRS shall be at the ground surface or at some specified depth in the soil column below the ground surface associated with the base of the foundation.

Basic information to characterize the site and to determine the ground motion shall be found by conducting a PSHA according to ANSI/ANS 2.27 and 2.29.

2.2 DEVELOPMENT OF DESIGN BASIS EARTHQUAKE GROUND MOTIONS

For Seismic Design Categories (SDC) 2, 3, 4, and 5, the UHRS for horizontal shaking shall be specified at two mean annual frequencies of exceedance, $H_P = P_F$, defined in Table 1-1, and $H_D = 10 \times P_F$, where H_D is used to define the slope of the hazard curve. The DBE ground motion for horizontal shaking shall be defined in terms of the design response spectra (DRS) given by

$$\mathrm{DRS} = \mathrm{SF} \times \mathrm{UHRS}_{H_P} \tag{2-1}$$

where UHRS_{H_P} is the UHRS determined for the exceedance frequency H_P, and SF is the scale factor, which is defined in Equation (2-3), at each spectral frequency.

For each spectral frequency at which the UHRS is defined, a slope factor, A_R, shall be determined from

$$A_R = \frac{\mathrm{SA}_{H_P}}{\mathrm{SA}_{H_D}} \tag{2-2}$$

where SA_{H_P} is the spectral acceleration (SA) at the exceedance frequency H_P, and SA_{H_D} is the spectral acceleration at the exceedance frequency H_D. The SF at this spectral frequency is given by

$$\mathrm{SF} = \mathrm{Maximum}\left[\mathrm{SF}_1, \mathrm{SF}_2, \mathrm{SF}_3\right] \tag{2-3a}$$

$$\mathrm{SF}_1 = A_R^{-1.0} \tag{2-3b}$$

$$\mathrm{SF}_2 = 0.6 A_R^{-0.2} \tag{2-3c}$$

$$\mathrm{SF}_3 = 0.45 \tag{2-3d}$$

The DRS is typically defined at the same control location in the free field as that at which the hazard curve and the UHRS are defined. Provisions are given in Section 2.3 for defining this DRS at other heights in the soil column.

If required, the DRS shall be amplitude scaled up by one factor across the entire frequency range such that the zero period acceleration is not less than 0.04 g for SDC 2, 0.06 g for SDC 3, 0.08 g for SDC 4, or 0.10 g for SDC 5.

2.3 METHOD TO DEFINE DESIGN RESPONSE SPECTRA AT VARIOUS DEPTHS IN THE SITE PROFILE

This section presents provisions for defining the DRS at other locations in the site profile besides the control location. The control location is typically defined at the bedrock outcrop. The free surface at the top of the soil profile is the most common location at which the DRS is to be determined. The DRS may also be determined at other locations in the profile following these procedures.

Before performing the site response evaluations, the characteristic earthquakes (magnitude and distances) at each spectral frequency at which the DRS is to be computed shall be obtained. The spectral shapes associated with these characteristic events shall be scaled to the bedrock outcrop UHRS at each spectral frequency of interest.

Convolution of the bedrock outcrop UHRS to the surface or other depths in the soil column must consider the following effects on the amplification of the bedrock input motion, site amplification functions (SAFs), as it propagates through the soil column:

1. Magnitude of characteristic earthquakes,
2. Distance of the characteristic earthquake from the site,
3. Nonlinearity of soil properties,
4. Uncertainty in the site profile, and
5. Uncertainty in the dynamic soil properties.

Simplifying assumptions as to the magnitude, distance, and frequency effects on the SAF are permitted if justified.

The approach for obtaining the DRS at the ground surface (or at some other intermediate depth) is summarized as follows:

1. Convolve the rock UHRS at the hazard mean annual frequency of exceedance, H_P, to obtain the corresponding UHRS_{H_P} at the ground surface (or other location in the soil column) using site-specific soil properties.
2. Convolve the rock UHRS at $H_D = 10H_P$ to obtain the corresponding UHRS_{H_D} at the ground surface (or other location in the soil column).
3. Determine the slope factor, A_R, from the ratio of SA_{H_P} to SA_{H_D} at the ground surface (or other location in the soil column), computed over the spectral frequency range, frequency by frequency, using Equation (2-2).
4. Use Equation (2-3a) to develop the scale factor, SF, at each spectral frequency, at the ground surface (or other location in the soil column).

5. Generate the DRS per Equation (2-1).

The number of convolution calculations performed must be sufficient to capture effects of the variability and uncertainty in soil properties on site response.

2.4 CRITERIA FOR DEVELOPING SYNTHETIC OR MODIFIED RECORDED ACCELERATION TIME SERIES

Ground motions that are generated to match or envelop given design response spectral shapes defined in Section 2.2 shall comply with Steps 1 through 6 in this section. The general objective is to generate a modified recorded or synthetic acceleration time series that achieves approximately a mean-based fit to the target spectrum, that is, the average ratio of the spectral acceleration calculated from the time series to the target, where the ratio is calculated frequency by frequency, is only slightly greater than one. A modified recorded acceleration time series is developed by scaling an earthquake (seed) record of appropriate magnitude and distance characteristics to match a given design response spectral shape. A synthetic acceleration time series is defined as a motion developed without the use of a real, recorded ground motion seed. Site-specific analyses shall use modified recorded acceleration time series, whereas site-independent analyses may use synthetic acceleration time series. The aim of either approach is to achieve a time series that does not have significant gaps in the Fourier amplitude spectrum but that is not biased high with respect to the target. Acceleration time series biased high with respect to a spectral target may overdrive (overestimate damping and stiffness reduction) a site soil column or structure when nonlinear effects are important.

1. The time series shall have a sufficiently small time step and sufficiently long duration. Records shall have a Nyquist frequency of at least 50 Hz (e.g., a time step of, at most, 0.010 s) and a total duration of at least 20 s. If frequencies higher than one-third of the Nyquist frequency are of interest, the time step of the record must be suitably reduced. Refinement of the time series shall be achieved by using the Fourier (frequency) domain technique with zero packing.
2. Spectral accelerations at 5% damping shall be computed at a minimum of 100 points per frequency decade, uniformly spaced over the log frequency scale. The comparison of the response spectrum of the time series with the target spectrum shall be made at each computed frequency point in the range from at least 0.2 Hz to the highest frequency of interest. The highest frequency of interest must exceed the frequency of the peak of the target response spectrum but be no less than 25 Hz.
3. The computed 5% damped response spectrum of the time series (if one set of three-component time series is used for analysis) or of the average of all time series (if a suite of motions is used for analysis) shall not fall more than 10% below the target spectrum at any one frequency. To prevent spectra in large frequency windows from falling below the target spectrum, the spectra within a frequency window of no larger than 10% centered on the frequency are allowed to fall below the target spectrum. This window corresponds to spectra at no more than nine adjacent frequency points defined in Step 2 from falling below the target spectrum.
4. The computed 5% damped response spectrum of the time series (if one set of three-component time series is used for analysis) or the mean of the 5% damped response spectra (if a suite of motions is used for analysis) in any direction shall not exceed the target spectrum at any frequency by more than 30% (a factor of 1.3). If the spectrum for the time series exceeds the target spectrum by more than 30% at any frequency, a target power spectral density (PSD) shall be developed and the PSD of the time series shall be computed and shown to be consistent with the target.
5. Site-specific acceleration time series defined earlier shall have strong motion durations (defined by the 5% to 75% Arias intensity), and ratios V/A and AD/V^2 (A, V, and D are the peak ground acceleration, velocity, and displacement, respectively) that are generally consistent with characteristic values for the magnitude and distance of the appropriate controlling events defined for the UHRS.
6. For each earthquake simulation, the three spatial components shall be statistically independent. This independence is achieved when the correlation coefficients between each pair of the three spatial components do not exceed a value of 0.16. If relatively uncoupled response of the structure is expected in the three directions of motion, then only one time series is required in each direction, with each motion satisfying the enveloping criteria to the target design spectrum in that direction. Then the seismic analysis for each direction may be performed separately, and the methods described can be used to combine the three directional in-structure responses.

Synthetic or modified recorded acceleration time series are permitted for linear seismic analyses. Modified recorded acceleration time series shall be used for nonlinear seismic analyses. However, to meet these requirements, many recorded earthquake motions may be required. A modified recorded acceleration series history is a record of acceleration versus time that has been produced from recorded acceleration history. However, the Fourier amplitudes of the seed record are scaled such that the resulting response spectrum envelops the target response spectrum in the manner described previously. The Fourier phasing in the modified recorded earthquake time series shall be similar to the phasing in the recorded seed earthquake time series.

The seed time series selected from earthquake records shall be selected to have their response spectra similar in shape to the target spectra across the frequency range of engineering interest to the seismic analysis. The selection of recorded seed acceleration time series is based on the identification of dominant magnitude and distance pairs and the site geology that affects the site DRS. If the target spectra include the contributions from multiple characteristic events, multiple time series shall be used in which individual time series are developed using Steps 1 through 6, with seed time series selected from earthquake records fairly representing the dominant magnitude/distance (M/D) events embodied in the target spectra in the frequency range of engineering interest to the seismic analysis. Note that if one set of three-component time series is developed for each dominant M/D bin event, the enveloping criteria in Step 3 shall apply, whereas the averaging criteria in Step 3 apply if a suite of motions is used for each dominant M/D bin event. Alternatively, the records may be selected to match the dominant M/D pairs at the peak velocity and acceleration segments of the design spectrum. If behavior at the peak displacement frequency range is of interest, additional records are to be selected whose controlling event is appropriate at these frequencies.

CHAPTER 3
EVALUATION OF SEISMIC DEMAND

3.1 INTRODUCTION

Seismic analysis shall be performed in accordance with the requirements of ASCE 4, and seismic demand shall be developed in accordance with this and other chapters of this standard. Seismic demands are combined with nonseismic demands in accordance with Chapter 5 of this standard.

Load paths for nonseismic and seismic loadings shall be identified. Components in these load paths shall be designed and detailed for the forces and displacements induced by nonseismic loadings and three translational components of earthquake shaking.

3.2 SEISMIC DEMAND DEVELOPMENT

The seismic demand on a structure, system, or component (SSC) is developed by applying analytical approaches in ASCE 4. ASCE 4 provides both deterministic and probabilistic analysis criteria. The seismic demand is then combined with the nonseismic demand using load combinations documented in Chapter 5 for design.

3.3 MODELING AND INPUT PARAMETERS

Modeling of SSCs for seismic analysis shall follow ASCE 4-16, Chapter 3. Criteria for modeling of mass (or effective weight) are provided in ASCE 4-16, Section 3.4. A discussion on modeling of coupled subsystems is provided in ASCE 4-16, Section 6.4. A discussion on modeling of distribution systems, including stiffness, is included in ASCE 4-16, Chapter 10.

3.3.1 Effective Stiffness of Reinforced Concrete and Steel-Plate Composite Wall Members. Table 3-2 in ASCE 4-16 enables calculation of effective stiffness of reinforced concrete. AISC (2012) provides demand for steel-plate composite (SC) and concrete elements.

3.3.2 Effective Seismic Weight: Effective seismic weight shall be calculated consistent with ASCE 4-16, Section 3.4.2.

3.3.3 Damping Values for Structures, Systems, and Components. The damping values to be used for determining seismic design loads by linear elastic analysis for SSCs are presented in ASCE 4-16, Tables 3-1 and 6-2 and are reproduced here in Table 3-1. These values are presented as a function of the average response level determined by demand-to-capacity ratios. The D_e/C ratios are calculated for structural members (where C = code capacity, and D_e = total elastically computed demand, including nonseismic loads). The appropriate response level can be estimated from Table 3-2. Values of damping for steel-plate composite (SC) members and heating, ventilation, and air conditioning (HVAC) are not addressed in ASCE 4-16 but are included in Table 3-1 in this standard.

When performing the seismic analysis, consideration of the actual response level is required for generation of in-structure response spectra. In lieu of iterative analyses to determine the actual response level and associated damping value, Response Level 1 damping values are permitted for generation of in-structure spectra. Response Level 1 damping values shall be used if elastic buckling or brittle material failure controls the design.

Response Level 3 damping may be used for evaluating seismic-induced forces and moments in structural members by elastic analysis without consideration of the actual response level for Limit States A, B, or C. Response Level 2 damping may be used for Limit State D.

Response Level 1 damping values shall be used when performing a nonlinear inelastic response analysis that explicitly incorporates hysteretic energy dissipation. A summary of the maximum response level that shall be used for selecting the damping value is shown in Table 3-3.

Table 3-1. Specified Damping Values for Dynamic Analysis.

Type of Component	Damping (% of Critical) Response Level 1	Response Level 2	Response Level 3
Welded and friction-bolted metal structures	2	4	7
Bearing-bolted metal structures	4	7	10
Prestressed concrete structures (without complete loss of prestress)	2	5	7
Reinforced concrete structures	4	7	10
Steel-plate composite (SC) walls	3	5	10
Reinforced masonry shear walls	4	7	10
Piping	5	5	5
HVAC:			
• Pocket lock	7	10	[a]
• Companion angle	5	7	[a]
• Welded	3	4	[a]
Distribution systems:			
• Cable trays 50% or more full and in-structure response spectrum zero period acceleration of 0.25 *g* or greater	5	10	15
• For other cable trays, cable trays with rigid fireproofing and conduits	5	7	7
Massive, low-stressed mechanical components (e.g., pumps, compressors, fans, and motors)	2	3	[a]
Light welded instrument racks	2	3	[a]
Electrical cabinets and other equipment	3	4	5[b]
Liquid containing metal tanks:			
• Impulsive mode	2	3	4
• Sloshing mode	0.5	0.5	0.5

[a] Should not be stressed to Response Level 3. Use damping for Response Level 2.

[b] May be used for anchorage and structural failure modes that are accompanied by at least some inelastic response. Response Level 1 damping values shall be used for functional failure modes such as relay chatter or relative displacement issues that may occur at a low cabinet stress level.

Table 3-2. Estimating Damping Response Level.

Response Level*	D_e/C
1	≤ 0.5
2	≈ 0.5 to 1.0
3	≥ 1.0

* See ASCE 4 (2017b), Chapter 3, for more information.

Table 3-3. Summary of Maximum Response Level for Damping.

Analysis	Maximum Response Level
Elastic buckling conditions control design	Response Level 1
Generation of in-structure spectra	Response Level 1 (Response Level 2, if justified)
Limit State D	Response Level 2
Limit States A, B, or C:	
Elastic analysis	Response Level 3*
Inelastic time-history response analysis	Response Level 1 (Response Level 2, if justified)

* Only to be used with adequate ductile detailing. However, functionality of SSCs must be given due consideration. Response Level 3 corresponds to Limit State C; Response Level 3 may also be used for Limit States A and B.

CHAPTER 4
STRUCTURAL CAPACITY

4.1 STRUCTURAL SYSTEMS

Seismic force-resisting systems in new nuclear facilities shall be classified as one of the types listed in Table 4-1.

A structure may have different seismic force-resisting systems (e.g., braced frame and shear wall) along each of its major horizontal dimensions.

Seismic force-resisting systems for nuclear facilities shall possess adequate strength and ductility to survive design basis earthquake shaking and to perform adequately in shaking more intense than design basis. Structural systems that must also provide confinement shall possess adequate stiffness to limit deformations such that cracking in those exterior components that provide confinement is minimized. Acceptable seismic force-resisting systems include those identified in Table 4-1. There are no height limits on the acceptable seismic force-resisting systems. Framing systems other than those listed in Table 4-1 may be acceptable for new nuclear facilities. The use of an undefined seismic force-resisting system shall be justified on a project-specific basis using nonlinear response-history analysis of three-dimensional framing systems composed of verified and validated models of structural components according to ASCE 4. Independent peer review of an undefined seismic force-resisting system is required.

Consensus standards shall be used for the design of the seismic force-resisting systems listed in Table 4-1, as supplemented by the requirements for seismic demand and capacity calculations, ductile detailing, peer review, and confinement (if required) that are specified in this standard and by additional project-specific needs.

Nonductile seismic force-resisting systems are prohibited for new nuclear facilities. Such systems include ordinary and intermediate reinforced concrete and steel moment-resisting frames; reinforced concrete moment-resisting frames composed of columns and flat slabs, flat plates, or waffle slabs; plain concrete frames and walls; unreinforced masonry walls; and timber frames and walls.

4.2 COMPONENT CAPACITIES

4.2.1 General. Force capacities shall be based on either (a) strength design (or load and resistance factor design) calculations, or (b) allowable stress (strength) calculations, as increased by a factor as described in this chapter. Deformation capacities are generally not calculated, but sufficient capacity is assumed if ductile details are used.

4.2.2 Reinforced Concrete. The design and detailing of reinforced concrete components shall be based on the provisions of ACI 349 (2013b).

4.2.2.1 Shear Strength of Reinforced Concrete Members. The shear strength provided by concrete for nonprestressed members, V_c, calculated in accordance with Section 11-2 of ACI 349-13 (2013b), shall be multiplied by a size correction factor κ. The size correction factor κ is a function of section effective depth, d, in inches, as follows:

Members with minimum shear reinforcement:

$$\kappa = 1 \tag{4-1a}$$

Members without minimum shear reinforcement:

$$\kappa = \frac{2.8\sqrt[4]{100\rho}}{\sqrt{d}}, \; d \text{ in inches} \tag{4-1b}$$

$$\kappa = \frac{14.1\sqrt[4]{100\rho}}{\sqrt{d}}, \; d \text{ in mm} \tag{4-1b.si}$$

where ρ is the longitudinal reinforcement ratio, and the effective depth, d, in Equation (4-1b) shall not be taken as less than 5 in. (127 mm).

4.2.2.2 Combined In-Plane and Out-of-Plane Shear in Slabs, Diaphragms, and Walls. The in-plane and out-of-plane shear forces in slabs, diaphragms, and walls shall be combined as follows:

$$\left(\frac{V_u}{\phi V_n}\right)^2_{\text{In-plane}} + \left(\frac{V_u}{\phi V_n}\right)^2_{\text{Out-of-plane}} \leq 1.0 \tag{4-2}$$

4.2.3 Structural Steel. Strength design shall be used for capacity calculations of structural steel components wherever possible. If capacity is established using a standard based on allowable stress design, the design strength shall be limited to no more than 1.5 times the allowable stress design capacity.

The capacity of a steel component shall be calculated using one of the following standards:

- Carbon steel components: AISC N690 (2018);
- Stainless steel components: ASCE 8 (2002); or
- Cold-formed carbon steel components: AISI S100 (2012).

The design and detailing of carbon steel components shall be based on AISC 341 (2016b) to ensure that ductile system response and inelastic deformation capacity are considered sufficient for shaking more intense than the design basis.

Table 4-1. Acceptable Seismic Force-Resisting Systems for New Nuclear Facilities.

Basic Structural System	Seismic Force-Resisting System
Bearing wall system	Shear walls: Reinforced concrete according to ACI 349 (2013a) Steel-plate composite (SC) according to AISC N690 (2018) Masonry with Category D detailing according to ACI 530 and 530.1 (2013b) Light steel-framed bearing walls with tension-only bracing Braced frames according to AISC N690 (2018) and AISC 341 (2016b): Steel eccentrically braced frame (EBF) Buckling restrained braced frame (BRBF) Special and ordinary steel concentrically braced frame (CBF)
Building frame system	Shear walls: Reinforced concrete according to ACI 349 (2013a) SC according to AISC N690 (2018) Masonry with Category D detailing according to ACI 530 and 530.1 (2013b) Braced frames according to AISC N690 (2018) and AISC 341 (2016b): Steel EBF BRBF Special and ordinary steel CBF
Moment-resisting frame system	Moment-resisting frame: Reinforced concrete according to ACI 349 (2013a) Reinforced concrete slab-wall Steel according to AISC N690 (2018) and AISC 341 (2016b)

4.2.4 Steel-Plate Composite. The design and detailing of structural walls composed of steel faceplates and infill concrete shall be based on AISC N690 (2018).

4.2.5 Reinforced Masonry. The design and detailing of reinforced masonry walls shall be based on the strength procedures of ACI 530 and ACI 530.1 (2013a).

CHAPTER 5
LOAD COMBINATIONS AND ACCEPTANCE CRITERIA FOR STRUCTURES

5.1 LOAD COMBINATIONS

5.1.1 General. A structure shall be proportioned and detailed to resist the effects of gravity loads, operating loads, earthquake loads, and applicable accident loads. Load combinations that include seismic loads shall be as defined in Section 5.1. Remaining load combinations shall be as defined in the project design criteria.

5.1.2 Seismic Load Combinations. Seismic load combinations for strength-based acceptance criteria shall consider the inelastic energy absorption factor, F_μ. Seismic load combinations for deformation-based acceptance criteria shall combine the unfactored effects of both seismic and nonseismic loads.

5.1.2.1 Seismic Load Combinations for Strength-Based Acceptance Criteria. For elastic analyses, the total demand acting on an element shall be the sum of nonseismic demand, D_{NS}, and seismic demand, D_S, according to the following load combination as appropriate:

1. For bending moment, in-plane shear, and axial load in pairs of diagonal braces, use

$$D = D_{NS} + \frac{D_S}{F_\mu} \tag{5-1a}$$

2. For other axial loads, other shear loads, and torsion, use

$$D = D_{NS} + \frac{D_S}{1.0} \tag{5-1b}$$

where

D = Total demand acting on an element.

D_{NS} = Nonseismic demand acting on an element. Nonseismic demand shall include the mean effects of dead, live, equipment, fluid, snow, and at-rest lateral soil loads. The nonseismic demands shall be consistent with the mass defined in Section 3.3.2 of this standard.

D_S = Calculated seismic response to the design basis earthquake using an elastic analysis approach discussed in Chapter 3 and the appropriate damping values from Table 3-2. For determination of load case D_S for concrete structures, use of the provisions of ACI 349, Section 9.2.10, is prohibited.

F_μ = Inelastic energy absorption factor for structural elements, as defined in Section 5.1.3.

5.1.2.2 Seismic Load Combinations for Deformation-Based Acceptance Criteria. The total demand acting on an element for use with displacement-based acceptance criteria shall be the sum of seismic demand, D_S, and nonseismic demand, D_{NS}, as combined with the following load combination:

$$D = D_{NS} + D_S \tag{5-2}$$

where D, D_{NS}, and D_S are as defined in Section 5.1.2.1.

This load combination is used for nonlinear seismic analyses. Equation (5-2) shall also be used to evaluate deformations in linear analyses.

5.1.3 Inelastic Energy Absorption Factor. F_μ is the inelastic energy absorption factor for structural elements. The system inelastic energy absorption factor is derived as follows:

1. Determine $F_{\mu C}$, the component inelastic force reduction factor described in Section 5.1.3.1.
2. Determine $F_{\mu 1}$, the modified inelastic energy absorption factor for soft or weak story effects as described in Section 5.1.3.2.
3. Determine $F_{\mu 2}$, the modified inelastic energy absorption factor for high frequency as described in Section 5.1.3.3.
4. Determine $F_{\mu 3}$, the modified inelastic energy absorption factor for ratcheting as described in Section 5.1.3.4, and
5. $F_\mu = F_{\mu 3}$

As these steps show, the modifications to the component inelastic force reduction factor are cumulative. As an alternative, the value of F_μ is permitted to be taken as 1.0.

5.1.3.1 Component Inelastic Energy Absorption Factor. Earthquake demand given by forces, moments, or stresses (i.e., not deformations) may be reduced by the inelastic energy absorption factor, $F_{\mu C}$, provided in Table 5-1, for the given limit state.

1. Linear interpolation shall be used to determine values of $F_{\mu C}$ for intermediate values of ℓ/h, f_v, and P in Table 5-1.
2. The inelastic energy absorption factors in Table 5-1 are based on the ductile detailing requirements of Chapter 6 and are not applicable for structures that do not meet the requirements of Chapter 6.
3. $F_{\mu C}$ for Limit State D shall be unity.
4. $F_{\mu C}$ for compression and shear in columns shall be unity.
5. Unless justified by an anchorage-specific evaluation, the inelastic factor $F_{\mu C}$ for anchorage shall be unity.
6. $F_{\mu C}$ for out-of-plane behavior of concrete slab-wall systems that meet the detailing requirements of Chapter 6 shall be taken from values for slab-wall moment frames.
7. $F_{\mu C}$ for collector elements shall be unity.

It is permissible to develop a project-specific system energy absorption factor, $F_{\mu C}$, that meets the performance goals of this standard for new or existing structures. The development of a

Table 5-1. Inelastic Energy Absorption Factor, $F_{\mu C}$.

Limit State		Reduction Factor, F_{μ}[a] LS A	LS B	LS C
Special reinforced concrete moment frames:				
Beams	$15 \leq \ell/h$	5.25	4.0	2.5
	$\ell/h \leq 10$	3.25	3.0	2.5
Columns[b]		2.0	1.75	1.5
Reinforced concrete shear walls and steel-plate composite walls, in-plane:				
Flexure-critical walls, $\frac{h_w}{\ell_w} \geq 2.0$	$6\sqrt{f'_c} < f_v$	2.25	2.0	1.75
	$f_v < 3\sqrt{f'_c}$	2.5	2.25	1.75
Shear-critical walls, $\frac{h_w}{\ell_w} < 2.0$		2.0	1.75	1.5
Special steel moment frames:				
Beams and columns[b] $P < 0.2\phi P_n$		5.25	3.5	2.5
Columns[b] $P = 0.3\phi P_n$		4.25	3.0	1.25
Columns[b] $P = 0.4\phi P_n$		3.25	2.25	1.25
Columns[b] $P = 0.5\phi P_n$		1.75	1.5	1.25
Columns[b] $P > 0.5\phi P_n$		1.0	1.0	1.0
Steel braced frames:				
Special concentric (bracing members)		4	3	2
Ordinary concentric (bracing members)		2.5	2.0	1.5
Eccentric:				
Shear-critical link beam		6	4	2
Flexure-critical link beam		5	3.5	2.5
Buckling-restrained		6	4	2
Chevron		2.5	2.0	1.5
Slab-wall moment frames:				
Roof slabs, floor slabs, beams, and walls of reinforced concrete	$15 \leq \ell/h$	2.5	2.25	2.0
	$\ell/h \leq 10$	2.25	2.0	2.0

Notes: f'_c = concrete compressive strength; f_v = average shear stress, which is equal to shear divided by the shear area; h_w = height of the entire shear wall or a segment of the shear wall considered; P = nominal axial load; ϕP_n = design axial strength; ℓ = span length; h = beam depth; and ℓ_w = length of shear wall or segment of shear wall considered in direction of shear force.

[a] $F_{\mu C}$ = 1.0 for Limit State D.

[b] $F_{\mu C}$ for compression and shear in columns shall be taken as 1.0.

project-specific system energy absorption factor shall be fully justified and subject to peer review.

5.1.3.2 Weak or Soft Story Inelastic Energy Absorption Factor. A weak story is defined as a story with a lateral strength that is less than 80% of the story above. The story strength is the total strength of seismic force-resisting elements sharing the story shear for the direction under consideration.

A soft story is defined as a story with a lateral stiffness that is less than 70% of that in the story above or less than 80% of the average stiffness of the three stories above.

The reduced inelastic energy absorption factor for a structure with a weak or soft story, $F_{\mu 1}$, shall be determined from Equations (5-3a) and (5-3b).

When a structure has either a weak story or soft story,

$$F_{\mu 1} = 1 + 2(F_{\mu C} - 1)\frac{n - k + 1}{n(n + 1)} \quad (5\text{-}3a)$$

If the structure does not have either a weak or soft story,

$$F_{\mu 1} = F_{\mu C} \quad (5\text{-}3b)$$

where

$F_{\mu C}$ = Component inelastic energy absorption factor described in Section 5.1.3.1,

n = Number of stories in the structure, and

k = Story level of the highest weak or soft story in the structure.

Judgment shall be used in determining the number of stories in facilities that have irregular geometry, discontinuous floor elevations, and partial height walls.

The adjustment for the effects of a weak or soft story shall be applied only to element seismic demands in the weak or soft story and all stories below. For stories above the weak or soft story, $F_{\mu 1}$ equal to $F_{\mu C}$ shall be used.

5.1.3.3 High-Frequency Inelastic Energy Absorption Factor. $F_{\mu 2}$ depends on the frequency of the predominant structural response.

1. For a structure with predominant structural response at a frequency greater than the amplified acceleration region of the design response spectrum (DRS), $F_{\mu 2}$, shall be determined from Equation (5-4):

$$F_{\mu 2} = F_{\mu 1}\left(\frac{SA(f_n)}{SA(f_e)}\right) \geq 1.0 \qquad (5\text{-}4)$$

where

$SA(f_n)$ = Spectral acceleration at the frequency of the predominant structural mode, f_n, and
$SA(f_e)$ = Spectral acceleration at an effective natural frequency for nonlinear seismic response, f_e, as defined by Equation (5-5).

if $f_n \leq f_{\text{PEAK SA}}$, then $f_e = f_n$.
if $f_n > f_{\text{PEAK SA}}$, then

$$f_e = f_n\sqrt{\frac{2}{F_{\mu 1}^2 + 1}} \geq f_{\text{PEAK SA}} \qquad (5\text{-}5)$$

where $f_{\text{PEAK SA}}$ is the upper frequency of the amplified acceleration region of the DRS.

2. For a structure with predominant structural response at a frequency less than or equal to the amplified acceleration region of the DRS, $F_{\mu 1}$, need not be reduced for high frequency.

5.1.3.4 Adjustments to the Inelastic Energy Absorption Factor for Ratcheting. The values of $F_{\mu C}$ from Table 5-1 as adjusted according to Sections 5.1.3.2 and 5.1.3.3 shall be further adjusted for components with potential for significant ratcheting, as follows:

$$F_{\mu 3} = \alpha_{F_\mu} \times (F_{\mu 2} - 1) + 1 \qquad (5\text{-}6)$$

where

$F_{\mu 2}$ = Inelastic energy absorption factor from Section 5.1.3.3; and
α_{F_μ} = Ratcheting modification coefficient. The factor is dependent on the ratio of nonseismic demand, D_{NS}, to code capacity, C, as follows:

$\alpha_{F_\mu} = 1$ for $D_{NS}/C = 0$;
$\alpha_{F_\mu} = 0.6$ for $0.1 \leq D_{NS}/C \leq 0.4$, and
$\alpha_{F_\mu} = 1$ for $D_{NS}/C \geq 0.6$.

Linear interpolation shall be used to determine values of α_{F_μ} for intermediate values of D_{NS}/C.

Exception: Alternative methods to account for ratcheting include the following:

1. Designing the ratcheting-sensitive element with $F_{\mu 3} = 1.0$, or
2. Computing $F_{\mu 3}$ using nonlinear analysis that accounts for both sustained nonseismic loads and seismic effects.

5.2 ACCEPTANCE CRITERIA

5.2.1 General. The applicable acceptance criteria depend on the type(s) of analysis performed for the design. Linear analyses shall meet the strength acceptance criteria in Section 5.2.2(a) for the load combination in Equations (5-1a) and (5-1b) and the displacement criteria in Section 5.2.3.1(a) for the load combination in Equation (5-2). Nonlinear analyses shall meet the displacement acceptance criteria in Sections 5.2.3.1 and 5.2.3.2. The capacity of yielding elements in nonlinear analyses shall meet the strength design criteria in Section 5.2.2(b). The load combination in Equation (5-2) shall be used in conjunction with the nonlinear acceptance criteria. For steel structures designed based on AISC allowable stress calculations, the allowable stresses shall be scaled to strength-based nominal code capacity according to Chapter 4.

5.2.2 Strength-Based Acceptance Criteria. For linear analyses, the total demand acting on an element, D, shall be less than or equal to the element's code capacity, C:

Table 5-2. Allowable Drift Ratio Limits as a Function of Limit State and Structural Systems.

	Allowable Drift Ratio Limit, γ_a			
Limit State	**LS A**	**LS B**	**LS C**	**LS D**
Special reinforced concrete moment frames	0.025	0.015	0.010	0.005
Reinforced concrete shear walls and steel-plate composite walls, in-plane:				
Flexure-controlled walls, $\frac{h_w}{\ell_w} \geq 2.0$				
$6\sqrt{f_c'} < f_v$	0.008	0.006	0.004	0.004
$f_v < 3\sqrt{f_c'}$	0.010	0.008	0.005	0.005
Shear-controlled walls, $\frac{h_w}{\ell_w} < 2.0$	0.008	0.006	0.004	0.004
Special steel moment frames	0.035	0.025	0.010	0.005
Steel braced frames:				
Concentric	0.020	0.013	0.005	0.005
Eccentric and buckling-restrained	0.030	0.017	0.005	0.005

Table 5-3. Allowable Rotation Limits for Nonlinear Analysis.

Limit State		Allowable Nonlinear Hinge Rotation θ_A* LS A	LS B	LS C
Special reinforced concrete moment frames:				
Beams	$15 \leq \ell/h$	0.020	0.010	0.005
	$\ell/h \leq 10$	0.010	0.008	0.005
Columns		0.005	0.003	0.000
Special steel moment frames:				
Beams and columns	$P < 0.2\,\phi P_n$	0.030	0.017	0.004
Columns	$P = 0.3\,\phi P_n$	0.021	0.012	0.004
Columns	$P = 0.4\,\phi P_n$	0.013	0.009	0.004
Columns	$P = 0.5\,\phi P_n$	0.006	0.005	0.004
Columns	$P > 0.5\,\phi P_n$	0.000	0.000	0.000
Slab-wall moment frames:				
Roof slabs, floor slabs, beams, and walls of reinforced concrete				
	$15 \leq \ell/h$	0.010	0.008	0.005
	$\ell/h \leq 10$	0.008	0.006	0.005

* $\theta_A = 0$ for Limit State D.

$$D \leq C \quad (5\text{-}7)$$

where the code capacity is developed in accordance with Chapter 4 and the demand is developed using Equations (5-1a) and (5-1b), with seismic demand determined from Chapter 3.

For nonlinear analyses, the capacity of all elements, including yielding elements, shall be limited to code capacities:

- ϕM_n for bending,
- ϕV_n for shear, and
- ϕP_n for axial loads.

where M_n, V_n, and P_n are the nominal code capacities in flexure, shear, and axial force, respectively.

5.2.3 Deformation-Based Acceptance Criteria

5.2.3.1 Allowable Drift Limits for Structural Systems. The total story drift ratio for each story shall be less than the allowable values given in Table 5-2 where the total story drift ratio is calculated with the demand in Equation (5-2) (drift ratio, γ, is the story displacement divided by the story height).

For nonlinear analysis of shear wall or braced-frame structures, the total drift ratio of a given element for each story shall be less than the allowable drift ratio:

$$\gamma_t \leq \gamma_a \quad (5\text{-}8)$$

where

γ_t = Total drift ratio of an element (in radians) caused by the demand in Equation (5-2), and
γ_a = Allowable drift ratio (in radians) given in Table 5-2.

Linear interpolation shall be used to determine the allowable drift limits for intermediate values of f_y in Table 5-2.

Allowable drift limits for out-of-plane behavior of concrete wall-slab systems shall be taken from values for slab-wall moment frames.

5.2.3.2 Allowable Rotation Limits for Structural Members. For nonlinear analysis of frame structures, the nonlinear hinge rotation in beams and columns shall be less than the allowable nonlinear hinge rotation:

$$\theta_N \leq \theta_A \quad (5\text{-}9)$$

where

θ_N = Nonlinear hinge rotation in an element (in radians) caused by the demand in Equation (5-2), and
θ_A = Allowable hinge rotation (in radians) given in Table 5-3.

Linear interpolation shall be used to determine deformation limits for intermediate values of ℓ/h and P in Table 5-3.

The allowable hinge rotation for frame members, θ_A, in Table 5-3 represents the nonlinear rotation of a plastic hinge. Columns with $\theta_A = 0$ shall be designed to behave elastically.

CHAPTER 6
DUCTILE DETAILING REQUIREMENTS

The goal of this chapter is to enable ductile response at the system level. Ductile system response is achieved by regular framing systems with complete load paths, coupled with details that enable substantial inelastic response at the component levels: ductile details.

Structural systems that can achieve ductile response under earthquake shaking have a number of attributes, including: (a) a complete load path with each connection in the path designed and detailed to develop the capacity of the weaker member being connected, (b) a framing system that is regular in stiffness and strength, with minimal or no plan and/or vertical irregularities, and (c) components supporting gravity loads that are not part of the seismic load path detailed to retain their strength for seismic displacements expected in design basis and beyond design basis loadings. Structures, systems, and components (SSCs) that span between isolated (individual) buildings shall be designed and detailed to accommodate design forces and expected inelastic deformation of design basis and beyond design basis loadings.

This chapter provides requirements for detailing components of structural systems in nuclear facilities permitted for use by Chapter 4.

6.1 STEEL STRUCTURES

All steel structures shall meet the minimum requirements of ANSI/AISC N690 (2018).

6.1.1 Moment Frames. Moment frames shall be detailed in accordance with special moment frame (SMF) provisions of ANSI/AISC 341 (2016c).

6.1.2 Braced Frames. Braced frames shall be detailed in accordance with the ANSI/AISC 341 (2016c) provisions for ordinary concentrically braced frames (OCBF), special concentrically braced frames (SCBF), eccentrically braced frames (EBF), or buckling-restrained braced frames (BRBF).

6.1.3 Steel-Plate Composite Shear Walls. Steel-plate composite (SC) shear walls shall meet the minimum requirements of ANSI/AISC N690 (2018).

6.1.4 Collectors. Collectors serve to connect elements of the lateral force-resisting system to provide a continuous seismic load path. Collectors shall be provided at locations where a lateral load-resisting element is not continuous across the width of the diaphragm. The deformation of a collector and its connections shall be calculated, and its effects on floor deformation, story drift, and the response of other structural and nonstructural elements shall be assessed.

6.1.5 Nearly Rigid Platforms and Supports. Platforms and supports designed for rigidity that (a) have seismic demand increased by a factor of 1.5 and (b) use an inelastic energy absorption factor of $F_\mu = 1$ for all members do not have to comply with the ductile detailing requirements of Chapter 6.

6.2 REINFORCED CONCRETE

All reinforced concrete structures shall meet the minimum requirements of ACI 349 (2013b).

6.2.1 General. Weak-axis bending in walls that are part of a lateral load-resisting system shall be designed using capacity procedures and shall be detailed to have a design shear strength that corresponds to 125% of the wall's nominal moment strength. Alternately, (1) the out-of-plane design shear demand for walls shall be increased by a factor of 1.5; and (2) the inelastic energy absorption factor, F_μ, for all members on this level shall be set equal to 1 for flexure and shear.

Combined shear wall and frame systems with more than 20% of the total lateral load at any elevation resisted by the frame system shall include:

a. Detailing per Section 6.2.2 for slab-wall systems, and
b. Detailing per ACI 349 (2013b), Chapter 21 for moment frames.

6.2.2 Slab-Wall Moment Frame Systems. Slabs and walls in slab-wall moment frames shall satisfy the following:

1. The positive out-of-plane moment strength at the face of the slab-wall joint shall be no less than one-half the negative moment strength at the face of the joint.
2. Reinforcement shall be fully developed at the face of the joint.
3. The slab-wall joint shall be designed using the strut and tie procedures of ACI 349 (2013b). The strength of compression struts in the slab-wall joint shall be calculated using $\beta_s = 0.60$.
4. Joints shall be reinforced according to Section 15.4.2 of ACI 318 (2014). Ties shall be anchored with seismic hooks according to ACI 318 (2014).

6.2.3 Requirements for Members Not Proportioned to Resist Forces Induced by Earthquake Motions. Beams, columns, and/or flat slabs not designed as part of the lateral seismic force-resisting system shall be proportioned to meet the requirements of Section 21.3 of ACI 318 (2008). The lateral load capacity of members proportioned to meet the requirements of Section 21.3 of ACI 318 (2008) shall be neglected when determining the forces acting on the lateral seismic force-resisting system.

6.2.4 Collectors. Collectors serve to connect elements of the lateral force-resisting system to provide a continuous seismic load path. Collectors shall be provided at all locations where a shear wall does not extend across the entire width of a diaphragm to provide a complete load path. The reinforcement in a collector shall be provided in addition to diaphragm and/or slab reinforcement. The deformation of a collector and its connections shall be calculated, and its effects on floor deformation, story drift, and the response of other structural and nonstructural elements shall be assessed.

6.2.5 Joints in Floor Slabs. Control joints or expansion joints with discontinuous reinforcing shall not be located in the seismic load path.

The in-plane shear capacity of a joint with a groove or intentionally reduced cross section shall be greater than the adjacent diaphragm shear capacity to maintain a complete load path.

6.3 ANCHORAGE

An anchor is a steel element either cast into concrete or post-installed into hardened concrete and used to transmit applied loads, including headed bolts, headed studs, expansion anchors, undercut anchors, grouted anchors, specialty inserts, and adhesive anchors. Design of anchors to attach an SSC to reinforced concrete shall conform to the following requirements:

1. Anchors shall be qualified, in the intended environment, to verify capacity in cracked reinforced concrete under simulated seismic loads.
2. Regardless of the intended environment, anchors qualified for use only in uncracked concrete are prohibited for SDC 3 through SDC 5 SSCs.
3. Only cast-in-place, undercut anchors or through bolts shall be used for anchorage of SDC 5 SSCs.
4. Design of cast-in and postinstalled mechanical anchors in reinforced concrete shall be performed according to ACI 349.
5. Design and installation of adhesive anchors in reinforced concrete shall be performed according to ACI 318 (2014).
6. Adhesive anchors intended for use in elevated temperature and/or radiation environments shall be qualified in those environments.

CHAPTER 7
SPECIAL CONSIDERATIONS

7.1 ROCKING AND SLIDING OF UNANCHORED BODIES

In general, it is preferable to anchor components to prevent rocking and sliding. However, unanchored components are acceptable as long as the provisions of this section are satisfied.

Sliding and rocking displacement demands shall be calculated in accordance with ASCE 4 (2017b).

The allowable capacity for sliding and rocking shall be taken as one-half the displacement, linear or angular, available before unacceptable contact or other unacceptable performance, such as overturning, falling, or reaching design limits of umbilical lines or attachments.

The effects of sliding and rocking need not be combined when they have each been conservatively calculated as acting separately.

7.2 BUILDING SLIDING AND OVERTURNING

7.2.1 Building Sliding. The sliding stability of a building may be checked using a static evaluation, comparing the lateral seismic forces with the static resistance:

$$\text{Static resistance} > 1.1 \times V_{\text{BaseShear}} \tag{7-1a}$$

Sliding stability is demonstrated by satisfying this condition. The base shear from linear analysis is determined using the approach given in Section 7.4.1. For foundations, the static resistance is given by

$$V_R = C + N \times \mu + P_u \tag{7-1b}$$

where

C = Effective cohesion force (effective cohesion stress, c, times foundation base area),
N = Normal (compressive) force,
μ = Coefficient of friction (static), and
P_u = Passive resistance.

If included, traction on the side of the foundation is calculated in a similar manner. The potential for reduced soil support of the structure shall be considered when accounting for the side traction. The coefficient of friction is determined from geotechnical investigations. In the absence of such information, μ is permitted to be based on the minimum of the effective internal friction angle of the soil and the friction coefficient between soil and foundation from published references. The passive pressure resistance and the associated deformation shall include internal friction and cohesion, as appropriate.

The horizontal displacement of the foundation is determined using the load-versus-displacement curve for the combined effect of cohesion, friction, and passive resistance. The functionality and structural integrity of the building and its appurtenances shall be checked, and if it is not impaired, an acceptable condition exists.

If Equation (7-1a) is not satisfied, it is permissible to demonstrate an acceptable condition by estimating the larger foundation movement resulting when the dynamic base shear momentarily exceeds the soil resistance and then demonstrating that the building remains functional for the resulting displacements. A refined estimate of the foundation movement shall recognize the time-dependent reversing nature of earthquake motion and the nonlinear soil resistance. If the functionality of the building is demonstrated for the larger displacements, an acceptable condition exists.

The nonlinear sliding analysis method of ASCE 4 (2017b) is permitted to estimate the foundation sliding movement, as long as the fundamental frequency, f_b, of the building exceeds the effective sliding frequency, f_{eS}, defined in Section 11.3.1 of ASCE 4 (2017b). If not, then substitute f_b for f_{eS} in the calculation of sliding displacement demand. When using this method, the appropriate coefficient of friction, μ, to be used in Section 11.3.1 of ASCE 4 (2017b) is given by

$$\mu = \frac{V_R}{W} \tag{7-2}$$

where V_R = total sliding resistance and W = building weight.

For a deeply embedded structure with a center of gravity below the site grade elevation on each perimeter wall, demonstration of sliding stability is not required.

7.2.2 Building Overturning. The acceptance criteria for overturning consist of a two-level check, which recognizes that the overturning moment from a fixed-base or linearly elastic soil-structure interaction (SSI) analysis is conservative. If the restoring moment exceeds the overturning moment (M_{OT}), a fixed-base or linearly elastic SSI analysis provides a reasonable estimate of the soil pressure. Therefore, building stability is demonstrated if

$$\text{Restoring moment} > 1.1 \times M_{OT} \tag{7-3}$$

and if the bearing pressure is acceptable. The judgment of acceptable bearing pressure shall ensure that displacements do not affect the functionality of the building and its appurtenances. The overturning moment from linear analysis is determined using the approach given in Section 5.1.2.1 of this standard.

For conditions where the M_{OT} from a fixed-base or linearly elastic SSI analysis exceeds the restoring moment, uplift is possible. The uplift condition reduces the contact area and increases the potential for soil failure. The building shall be evaluated for the uplift condition because there is considerable margin beyond Equation (7-3). The evaluation of the uplift condition shall include the appropriate overturning moment, soil pressure, building movement resulting from the uplift, effect of the building movement on buried piping, connections between buildings, and other items influenced by the rocking motion. The soil capacity for this evaluation shall be based on ultimate conditions and not limited by settlement criteria. If the associated structures, systems, and components (SSCs) remain functional, the overturning condition is acceptable.

The nonlinear rocking analysis method of ASCE 4 (2017b) is permitted to be used to demonstrate building stability and for estimating the maximum rocking angle, as long as the effective rocking frequency, f_e, computed from Section 11.3.2 of ASCE 4 (2017b) is less than the fundamental frequency of the building on soil with no uplift. When using this method, the building shall be assumed to rock about a point, B, located at the centroid of the soil contact area, A_C, given by

$$A_c = \frac{W}{q_{\text{ult}}} \quad (7\text{-}4)$$

where W = building weight and q_{ult} = ultimate bearing capacity of the soil not limited by settlement considerations.

For a deeply embedded structure with a center of gravity below the site grade elevation on each perimeter wall, demonstration of overturning stability is not required.

7.3 SEISMIC SEPARATION

A minimum separation shall be maintained between adjacent SSCs to accommodate relative displacements. The relative displacement, δ, shall be calculated as follows:

$$\delta = 2.0(\delta_1{}^2 + \delta_2{}^2)^{1/2} \quad (7\text{-}5)$$

where δ_1 and δ_2 are the displacements under design basis earthquake (DBE) loading of SSC_1 and SSC_2, respectively, calculated at any points of potential mutual pounding in the direction of a line connecting the points. Displacements δ_1 and δ_2 may be based on either linearly elastic or nonlinear analysis and shall include foundation and/or soil deformations if SSC_1 and SSC_2 are not founded on a common foundation. If δ_1 or δ_2 is calculated as the linearly elastic displacement of an inelastically designed (Limit State A, B, or C) structure with a fundamental frequency greater than 1 Hz, the calculated displacement shall be increased to account for the higher ratio of inelastic to elastic displacement associated in such structures.

Exception: The factor of 2 in Equation (7-5) is permitted to be reduced, but to no less than 1, if δ_1 and δ_2 are based on nonlinear response history analysis of the soil-structure system and the performance goals of (a) less than a 1% probability of impact in the DBE shaking and (b) less than a 10% probability of impact for the 150% DBE shaking, are achieved.

Similarly, when the combined response of independently displacing bodies affects acceptable performance, the independent displacements shall be combined by square root of the sum of the squares (SRSS).

7.4 SEISMIC DESIGN CONSIDERATIONS FOR FOUNDATION ELEMENTS

This section applies to deep foundation elements not addressed in ASCE 4 (2017b). Foundation elements consist of the supporting soil and structural elements to which the structure is attached. A continuous load path, or paths, through structural members with adequate strength and stiffness shall be provided to transfer all forces from the point of application to the supporting soil and/or pile foundation. The foundation shall be designed to resist the forces developed and to accommodate the movements imparted to the structure by the design ground motions. The soil capacity defined in the geotechnical report shall be used for the structural design of foundation elements subjected to combined seismic and nonseismic forces.

7.4.1 Linear Analyses. The seismic load on the supporting soil and/or pile foundation and connections of the piles and caissons to structural components is the load D_s described in Section 5.1.2.1 of this standard.

7.4.2 Nonlinear Analyses. The seismic distortions on (1) the foundation components and (2) the connections of piles and caissons to structural components correspond to the term D_s described in Section 5.1.2.2 of this standard. The force-displacement behavior of the foundation elements shall be calculated using the stiffness and strength obtained using established principles of soil mechanics. Variability of stiffness and strength must be considered in evaluating foundation demands to ensure estimates of demand on the foundation element at approximately the 80th percentile of nonexceedance probability.

7.4.3 Special Provisions for Foundation Components. Piling shall be designed and constructed to withstand maximum imposed curvatures from earthquake ground motions and structure response. Curvatures shall include free-field soil strains modified for soil–pile–structure interaction coupled with pile deformations induced by lateral pile resistance to seismic forces imposed by the structure.

7.4.4 Liquefaction Potential and Soil Strength Loss. The potential consequences to the structure of liquefaction or any soil strength loss shall be evaluated to ensure that the performance goals of this standard are met.

7.5 UNREINFORCED MASONRY WALLS USED FOR PARTITIONS, BARRIERS, OR RADIATION SHIELDING

Unreinforced masonry walls may be used for partitions, barriers or radiation shielding, and shall be designed per Section 1.5, and Appendices A and B of this standard.

7.6 PROVISIONS FOR CONSTRUCTION EFFECTS

Large and heavy nuclear structures with foundation basemat and/or sidewalls in contact with soil and backfill materials have potential to be subjected to substantial additional bending effects because of differential settlements caused by the heavy loads, soft or nonuniform site conditions, site excavation effects, anticipated groundwater variation during construction, or construction sequence effects. If a geotechnical evaluation indicates that the induced settlements have the potential to be large (greater than several inches [centimeters] across the site), the potential effect of such settlements on bending in the foundation elements shall be evaluated. Reinforcement design and detailing shall consider the potential effects of such differential settlements and shall minimize the potential for induced cracking of the foundation elements (walls and basemat).

Embedded walls shall be designed to take into account anticipated soil pressures expected during construction and normal operating and seismic conditions. Under normal operating conditions, lateral wall pressures may be calculated as the sum of (a) at-rest soil pressures, (b) effects of compaction of adjacent soils (if any), and (c) pressures induced by adjacent structures (if any). Dynamic pressures from seismic effects may be evaluated following the elastic procedures described in ASCE 4 (2017b). Additional dynamic pressures from adjacent structures, if any, shall be incorporated. If passive pressure assumptions are used in the evaluation of potential sliding of the embedded structure, the walls shall also be evaluated to ensure that they can sustain such pressures.

CHAPTER 8
SEISMIC QUALIFICATION OF EQUIPMENT AND DISTRIBUTION SYSTEMS

8.1 INTRODUCTION

Seismic qualification shall be based on equivalent static or dynamic analysis, testing, test experience data, earthquake experience data, or a combination of these methods. Permanent equipment and distribution systems shall be adequately supported and anchored to the structure. Whenever any method or a combination of these methods is chosen, it shall be demonstrated that the equipment or distribution systems are capable of performing all of their specified safety functions. It shall also be verified that the equipment and distribution systems do not interfere with the safety function of adjacent equipment or distribution systems according to ANS 2.26. The seismic demand for equipment and distribution systems shall be determined following the applicable methods and provisions of Chapter 3 of this standard, which includes considering three spatial components of input motion (two horizontal directions and the vertical direction). In some cases, additional input in the form of rotations about any (or all) of the three axes may be required.

Equipment and distribution systems may be supported by only their foundation (e.g., freestanding tanks in the yard), supported by the superstructure (e.g., supported directly on the main building floor slabs, walls, or frames), or a combination of the superstructure and a substructure that provides direct support to the equipment and distribution systems. The three cases are referred to herein as supporting structures.

8.2 QUALIFICATION BY ANALYSIS

Qualification of active electrical equipment by analysis shall not be permitted. Qualification of active mechanical equipment by analysis is permitted if the component is such that the functionality during an earthquake is established and a margin of loss of functionality during an earthquake is quantified. Qualification of active mechanical components by analysis shall be justified.

8.2.1 Seismic Analysis Methods. Seismic qualification by analysis shall use either an equivalent static analysis or a dynamic analysis. The following two sections provide the minimum requirements for each approach. The development of seismic demand shall satisfy Chapter 3 of this standard.

8.2.1.1 Equivalent Static Analysis. In equivalent static analysis, seismic inertial loads are applied statically as external loads, typically as body force accelerations, on the masses of the equipment and distribution systems as represented in an analytical model that includes the supports and anchorage. The results of the analysis are the resultant primary (equilibrium) stresses and the displacements. Equivalent static loads shall be developed from the in-structure response spectra at the location of the item considered, at the damping value according to Section 8.2.2.1, as appropriate for the seismic design category (SDC) of that equipment or distribution system. Analytical procedures for equipment and distribution systems shall meet the requirements of Section 4.5, "Equivalent Static Analysis," of ASCE 4 (2017b). For equivalent static analysis of piping systems, the method described in paragraph N-1225, "Simplified Dynamic Analysis," of the ASME Boiler and Pressure Vessel Code (BPVC), Section III, Division 1, Appendix N (ASME 2019c), is acceptable.

For equipment or distribution systems with multiple support points, support displacements caused by earthquake motion, termed seismic anchor motion (SAM), shall be considered as required in Section 4.3.5 of ASCE 4 (2017b) and in Paragraph N-1225 of ASME BPVC, Section III, Division 1, Appendix N (ASME 2019c), for piping systems.

8.2.1.2 Dynamic Analysis. For dynamic analysis, seismic loads and resultant inertia and SAM stresses in equipment and distribution systems shall be based on the earthquake dynamic input (i.e., in-structure response spectra or time histories) appropriate for the SDC of the equipment or distribution system. The dynamic analysis procedures specified in the following references shall be used as appropriate for the type of component:

- Appendix N of Section III, Division 1, of ASME BPVC (ASME 2019c);
- ASCE 4 (2017b);
- IEEE 344 (2013); and
- ASME QME-1 (2017a).

8.2.2 Demand for Qualification by Analysis. The following subsections provide the requirements to define demand on equipment and distribution systems when using qualification by analysis.

8.2.2.1 Damping. Damping values specified in Table 3-1 of Section 3.3.3 of this standard shall be used. Damping as a percentage of critical is defined for the type of component and the response level, which is defined in Table 3-3 of Section 3.3.3.

8.2.2.2 Inelastic Energy Absorption Factor. The inelastic energy absorption factors for equipment and distribution systems allowed for ductile-type behavior under seismic loads are provided in Table 8-1. If the component contains brittle material or joints in the load path, or if brittle material or joints are used that could affect its specified safety function, then F_μ values shall be taken as 1.0. To avoid overestimating energy dissipation, the inelastic energy absorption factor shall not be used with models that explicitly include inelastic energy absorption mechanisms unless justified.

8.2.2.3 Allowable Limit States for Active Mechanical Components and Pressure-Retaining Equipment. When analysis is used to qualify active mechanical components that change state as part of their safety function either during the earthquake shaking or immediately thereafter (i.e., seconds or minutes after the shaking stops), only Limit State D is permitted. Limit States D or C are permitted for active components if their change in state occurs following the earthquake by a period of time sufficient to determine if there is any damage to the plant or as necessary to reset tripped devices, such as relays. Limit State A is not permitted for pressure-retaining mechanical equipment that must retain leak-tight integrity.

8.2.2.4 Total Demand for Qualification by Analysis. The total demand on equipment and distribution systems qualified by analysis is given by the following equation:

$$D = D_{NS} + \frac{D_S}{F_\mu} \tag{8-1}$$

where

D = Total demand acting on the component,

D_{NS} = Nonseismic demand acting on the component (e.g., dead weight and fluid pressure). D_{NS} shall be consistent with the load combinations defined in the industry standards applicable to the capacity definition of Section 8.2.3.1, and

D_S = Seismic demand (response) of the component obtained by equivalent static or dynamic analysis. The input to the component analysis is generated from the analysis of the supporting structure(s) subjected to the design basis earthquake (DBE) and based on the SDC of the component.

If a dynamic analysis is performed and the input to the component is acceleration based, then an elastic analysis of the supporting structure(s) shall be performed to generate the input to the component, with F_μ of the supporting structure(s) set to 1.0.

If the input to the component is displacement based and the limit state(s) of the supporting structure(s) are other than Limit State D, then it may be nonconservative to assume that $F_\mu = 1.0$ for the supporting structure(s) for the analysis of the equipment and distribution systems. In this case, appropriate values of F_μ of the supporting structure(s) shall be taken into account.

The calculated seismic response of the component, D_S, shall be based on its effective natural frequency, f_e, as defined in Equation (5-5), where $F_{\mu 1}$ is equal to F_μ of the component (from Table 8-1), f_n is the predominant frequency of the component, and $f_{PEAK\ SA}$ is the peak of the in-structure response spectrum (ISRS) representing the component demand; and on its damping value according to Section 8.2.2.1.

When displacements need to be calculated as part of the seismic qualification by analysis, the demand given in Equation (8-1) shall be calculated with F_μ as described in the preceding paragraph.

8.2.3 Capacity Using Qualification by Analysis. The following sections define the capacity, C, of equipment and distribution systems when seismic qualification is based on analysis methods.

8.2.3.1 Capacity Defined by Industry Standards. Capacity shall be based on the stress or load limits in the appropriate code, standard, or specification for the equipment or distribution systems being qualified. If the capacity limits of ASME nuclear codes or standards (2019a, 2017a) are used to define capacity, then all ASME nuclear code construction requirements shall be met (an exception is permitted to be taken for nuclear grade stamping, when justified).

Table 8-1. Equipment and Distribution Systems Inelastic Energy Absorption Factor, F_μ.[a]

	Factor, F_μ		
	Limit State A[b]	Limit State B[b]	Limit State C[b]
Equipment			
Vessel[c]	1.50	1.25	1.15
Heat exchanger	1.50	1.25	1.15
Coolers	1.50	1.25	1.15
Chillers	1.50	1.25	1.15
Tanks[c] (vertical)	1.25	1.25	1.15
Tanks[c] (horizontal)	1.50	1.25	1.15
Pumps	1.50	1.25	1.15
Fans	1.50	1.25	1.15
Valves	1.50	1.25	1.15
Dampers	1.50	1.25	1.15
Filters[c]	2.00	1.50	1.25
Glove boxes[d]	2.00	1.50	1.15
Electrical boards[d]	2.00	1.50	1.15
Electrical racks[d]	2.00	1.50	1.15
Electrical cabinets[d]	2.00	1.50	1.15
Distribution systems			
Butt-joined groove welded pipe	1.75	1.50	1.25
Socket welded pipe	1.50	1.25	1.15
Threaded pipe	1.25	1.15	1.00
Conduit	1.50	1.35	1.25
Instrument tubing	1.50	1.35	1.25
Cable trays	1.50	1.35	1.25
HVAC ductwork	1.50	1.25	1.15
Equipment supports[c,d]	2.00	1.50	1.25

[a] These inelastic energy absorption factors, F_μ, are applicable to equipment functioning in a passive mode. For active components, the limit state is restricted to Limit State D, where $F_\mu = 1$. See Section 8.2.2.2.

[b] Except as discussed in notes [c] and [d], the allowable behavior limits for passive components are based on the ASME BPVC, Section III (2019b) allowables for Service Level D. It should be noted that ASME BPVC, Section III, Service Level D allowable stresses range from 1.6 to 2.0 times the yield stress.

[c] If the vessel, tank, or equipment supports are subject to elastic buckling, F_μ shall be taken as 1.0, regardless of the limit state.

[d] These components are normally designed to AISC capacities (2011), which are typically limited to 0.8 to 1.0 times the yield stress; hence, they are allowed a somewhat higher inelastic energy absorption factor as compared to ASME BPVC allowables (2019b), where allowable stresses can be as high as twice the yield stress.

8.2.3.2 Acceptance Criteria for Capacity. Demand for components, supports, and anchorage for active components shall be limited to less than the allowable capacity in the applicable code, standard, or specification. Where ASME BPVC, Section III, is specified, the ASME BPVC, Section III, Service Level D is allowed only for passive components limited to Limit States A or B. For active components, the ASME BPVC, Section III Service Level specified shall be justified to ensure the required safety function of the component.

8.2.4 Acceptance Criteria and Documentation for Qualification by Analysis. Acceptance criteria for qualification by analysis shall demonstrate that the capacity, *C*, exceeds the total demand, *D*, as follows:

$$D < C \quad (8\text{-}2)$$

Documentation shall meet the requirements of the code, standard, or specification used.

8.3 QUALIFICATION BY TESTING AND EXPERIENCE DATA

8.3.1 Tests and Experience Methods. The following sections provide the minimum general requirements, acceptance criteria, and governing references for qualification by testing and by use of experience data. In the following sections, *testing* refers to a component-specific shake table test, and *test experience data* refers to a database of test results for specific classes of equipment.

8.3.1.1 Testing. Seismic qualification by testing shall use either a proof test or a fragility test. A seismic proof test is performed using seismic input typically expressed as a test response spectrum (TRS). The TRS exceeds the required seismic input, usually defined as a required response spectrum (RRS). Seismic fragility testing is performed in steps of increased levels of seismic input until a failure or malfunction occurs or until the test limits of the shake table are reached. In either case, the test specimen shall be mounted on the test table (i.e., shake table) consistent with the mounting configuration in the nuclear facility. Any other nonseismic loads that exist concurrently with the seismic loading and that could affect its safety function shall be accounted for in the test program. Seismic qualification by testing shall meet the requirements of IEEE 344 (2013) or ASME QME-1 (2017a), depending on the type of item being tested. Seismic testing of in-line components, such as valve actuators, shall meet IEEE 382 (2006).

8.3.1.2 Test Experience Data. Available seismic test data on a specific class of equipment can be gathered, evaluated, and consolidated. These test data can be reduced to a test experience spectrum (TES), which defines the seismic response acceleration levels below which the class of equipment is expected to function. The TES is acceptable if all the specific equipment class caveats and inclusion rules are met and if it is documented that there are no design or material changes that can reduce the seismic ruggedness of the candidate equipment item when compared to those items tested that form the TES. The application of test experience methodology for seismic qualification in accordance with IEEE 344 (2013) is acceptable.

TES for chatter-sensitive devices (e.g., relays) shall be used to qualify the device for function during earthquake shaking only when sufficient verification of the device being represented (included) in the TES is presented.

8.3.1.3 Earthquake Experience Data. The procedures associated with verifying seismic adequacy using earthquake experience as given in ASME QME-1 (2017a) (Section QR and Appendix QR-A) are acceptable. Application of the earthquake experience methodology for seismic qualification specified in IEEE 344 (2013) is also acceptable. These procedures specify caveats and inclusion rules that shall be satisfied.

The application of earthquake experience, including operability, does not eliminate or substitute leak-tight and integrity requirements of ASME BPVC or B31 codes. Moreover, qualification of anchorage of the component shall be performed separately from application of experience-based approaches.

Items of newer vintage shall be investigated to ensure that they do not have any features that would reduce their seismic ruggedness as defined by the seismic experience data. If the candidate equipment item has complex features and its design varies significantly with time, then other methods of seismic qualification shall be used.

8.3.2 Demand for Qualification by Testing and Experience Data

8.3.2.1 Demand for Qualification by Testing. The demand for qualification by testing is the following:

$$D = D_{NS} + \gamma_{\text{test}}\, D_S \quad (8\text{-}3)$$

where

D = Total demand on the component;

D_{NS} = Additional loads such as pressure or state of the equipment (e.g., energized electrical panels or devices) that are expected to be present during the earthquake shaking are included in D_{NS}. Also, D_{NS} could be any preaging required before the seismic test. Often, $D_{NS} = 0$, other than the dead weight of components; and

D_S = Seismic demand (response) imposed on the component. Seismic demands are most often defined by a required response spectrum (RRS) or acceleration time histories, generated from the analysis of the supporting structure(s) subjected to the DBE and based on the SDC of the component. If the input to the component is acceleration based, then an elastic analysis of the supporting structures shall be performed to generate the input to the component, with F_μ of the supporting structures set to 1.0.

If the input to the component is displacement based and the limit state(s) of the supporting structure(s) are other than Limit State D, then it may be nonconservative to assume that $F_\mu = 1.0$ for the supporting structure(s) for the qualification of equipment and distribution systems. In this case, appropriate values of F_μ of the supporting structure(s) shall be taken into account when defining the displacement states imposed during qualification. The calculated seismic response of the component, D_S, shall be based on its effective natural frequency, f_e, as defined in Equation (5-5), where $F_{\mu 1}$ is equal to F_μ of the component (from Table 8-1), f_n is the predominant frequency of the component, and $f_{\text{PEAK SA}}$ is the peak of the ISRS representing the component demand, and on its damping value according to Section 8.2.2.1.

When displacements need to be calculated by analysis, the demand given in Equation (8-3) shall be calculated with F_μ as described in the preceding paragraph.

The factor γ_{test} is the ratio of the TRS to the RRS for qualification by testing. If the qualification testing is performed in accordance with IEEE 344, setting the factor γ_{test} to 1.33 meets the performance goals of this standard. If testing is performed to other procedures and criteria, the value of γ_{test} shall be determined so as to meet the performance goals of this standard.

There is no inelastic energy absorption factor (F_μ) in Equation (8-3) because the limit state is indirectly specified for the component (e.g., functional requirements or structural integrity) that defines its acceptance criteria.

8.3.2.2 Demand for Qualification by Test Experience Data and Earthquake Experience Data. Demand for qualification by test experience data is also given by Equation (8-3) with γ_{test} replaced by γ_{TES}, where γ_{TES} is the factor to be applied to the text experience spectrum (TES) required to meet the performance goals of this standard. For nonrelay generic equipment ruggedness spectra (GERS), setting γ_{TES} to 1.33 meets the performance goals of this standard. For relay GERS, setting γ_{TES} to 1.75 meets the performance goals of this standard. For any component using seismic input other than GERS, γ_{TES} shall be determined so as to meet the performance goals of this standard.

The demand for qualification by earthquake experience data (EED) is given by the following equation:

$$D = \gamma_{EED} D_S \tag{8-4}$$

where D and D_S are defined in Section 8.3.2.1. The factor γ_{EED} is the factor to be applied to earthquake experience data to meet the performance goals of this standard. The seismic demand from earthquake experience data is defined as an earthquake experience spectrum (EES). If the seismic demand D_S is defined by a Seismic Qualification Utility Group (SQUG) reference spectrum, which is an example of an EES, setting γ_{EED} to 1.0 meets the performance goals of this standard. If the earthquake experience data are based on other than a SQUG reference spectrum, the factor γ_{EED} shall be determined so as to meet the performance goals of this standard.

Earthquake experience data address equipment earthquake performance under normal operating conditions, as seen in industrial facilities. Nonseismic normal operating loads are naturally present in the earthquake experience data for the equipment and need not be accounted for in a separate term in Equation (8-4). If the candidate equipment item has atypical nonseismic loads, the earthquake experience method of qualification shall not be used unless justification is provided that the atypical load condition does not affect seismic ruggedness or function.

8.3.3 Capacity Defined for Seismic Qualification by Testing and Experience Data. Capacity, C, based on testing is defined as the applied test motion expressed in the frequency domain by the TRS for which it is demonstrated that the component maintains its required function. The specific requirements concerning test capacity shall satisfy the latest revision of IEEE 344 or ASME QME-1, as appropriate for the component being qualified.

Capacity, C, based on test experience data is defined by the TES developed by the equipment seismic test data for the specific and limited set of equipment that was tested. Capacity, C, based on earthquake experience data is defined as an EES, which is a definition of the equipment capacity based on earthquake ground motion at the database sites used to develop an equipment class. The TES or EES is acceptable for use if the component meets the equipment class caveats and inclusion rules and design changes (vintage) are appropriately evaluated to show no reduction in seismic ruggedness.

8.3.4 Acceptance Criteria and Documentation for Qualification by Tests and Experience Data. Acceptance criteria for qualification by test or experience data are defined by Equation (8-2). Acceptability of capacity not exceeding demand at a particular frequency or range of frequencies shall satisfy IEEE 344 or ASME QME-1, as appropriate for the type of component or method of qualification. The specific functionality of the component shall be demonstrated by the qualification method used.

Documentation shall meet the requirements of the code, standard, or procedure used and shall clearly indicate that capacity is greater than demand for the specified functional requirements.

CHAPTER 9
SEISMICALLY ISOLATED STRUCTURES

9.1 INTRODUCTION

Seismically isolated, safety-related nuclear structures shall be designed and constructed in accordance with this chapter and applicable requirements of this standard.

Chapter 9 is written for the seismic isolation of nuclear power plants and other nuclear structures. Application to the isolation of components of power reactors, small modular power reactors, or other components of safety-related nuclear structures, such as emergency diesel generators, is not addressed in detail in this chapter but is not precluded by these provisions.

Seismic isolation is provided in the horizontal plane only. These provisions do not address vertical isolation.

The provisions assume that the isolation system is protected from other hazards, such as flooding, so that it can perform its intended safety function.

Table 12-1 of ASCE 4 (2017b) summarizes the performance expectations for seismically isolated nuclear structures. Analysis shall be performed for design basis earthquake (DBE) shaking and 150% of DBE shaking, which is defined as beyond design basis earthquake (BDBE) shaking in this chapter. Seismic demands shall be calculated at the 80th percentile level for DBE shaking and the 90th percentile level for BDBE shaking. It shall be demonstrated by analysis that the isolated superstructure achieves the performance criteria of Section 1.3 of this standard.

A stop or displacement restraint shall be provided along each horizontal axis of the isolated structure at no less than the 90th percentile level for BDBE displacement along that axis.

Isolators and isolation systems shall be qualified by the following:

1. Dynamic testing of full-scale (prototype) isolators for compressive and tensile axial loads and bidirectional horizontal motion at amplitudes of displacement expected for BDBE ground motions;
2. Development of verified and validated numerical models capable of predicting the results of dynamic testing of prototype isolators, including deterioration of hysteresis caused by energy dissipation during earthquakes;
3. Demonstration through basic chemistry, laboratory tests, and field applications that the mechanical properties of the isolators do not change by more than 20% over a 50- to 100-year period in the temperature range of 40 °F (4.4 °C) to 80 °F (26.67 °C);
4. System-level testing of the isolation system using three translational components of earthquake ground motion; and
5. Verification and validation of numerical tools and codes to predict the seismic response of the isolation system.

Experienced persons who are independent of the isolator manufacturer shall perform these tasks. Results shall be peer-reviewed.

Hybrid isolation systems involving different types of bearing shall not be used.

Acceptable methods for dynamic analysis of a seismically isolated nuclear structure are presented in Chapter 12 of ASCE 4 (2017b). Each method involves soil–structure interaction analysis, with ground motion defined and input at the boundaries of the mathematical model of the soil–structure system. Three translational components of ground motion shall be used for analysis.

The provisions assume that the basemat immediately above the isolation system and the foundation below the isolation system are stiff in their plane and out of their plane so as to engage all isolators in the system in resisting gravity and earthquake-induced load effects. Requirements are provided in Chapter 12 of ASCE 4 (2017b).

9.2 GENERAL REQUIREMENTS

9.2.1 Isolation System

9.2.1.1 General. The design of the seismic isolation system shall account for the effects of gravity loads, lateral loads induced by earthquakes and wind, and for other effects as appropriate, including aging, creep, operating temperature, and moisture.

9.2.1.2 Vertical and Horizontal Load Resistance. Each element of the isolation system shall be designed to suffer no permanent damage under vertical loads and horizontal displacements associated with DBE shaking. The horizontal displacement of the isolation system, D_D, shall be calculated at the center of mass of the isolated superstructure, projected on the basemat, at the 80th percentile level.

Each element of the isolation system shall be designed to resist vertical loads and horizontal forces associated with a displacement equal to the clearance to the stop, CS.

9.2.1.3 Minimum Lateral Restoring Force. The horizontal force-displacement relationship of the isolation system shall be such that the force at the horizontal displacement D_D is greater than the force at horizontal displacement $0.5D_D$ by at least $0.025W$.

9.2.1.4 Wind Loads. Seismically isolated nuclear structures shall resist the effects of design wind loading in accordance with standards for nonisolated nuclear structures. The lateral displacement of the center of mass of the superstructure under design wind loadings should either (a) result in shear strain no greater than 10% in a low-damping (natural) rubber (LDR) bearing, or (b) be less than the yield displacement for lead (natural) rubber (LR) and friction pendulum (FP) bearings.

9.2.1.5 Operating Conditions. The isolation system shall be maintained at a temperature between 40 °F (4.4 °C) to 80 °F (26.67 °C);

Exception: A greater range of operating temperature is acceptable if it can be shown that the mechanical properties of the isolation system do not vary by more than ±20% from the values used for analysis and design, with 95% probability, over the life span of the structure.

9.2.1.6 Inspection and Replacement. The design shall provide access for inspection and replacement of all components of the isolation system.

The responsible design professional shall develop programs for monitoring the isolation system and inspection of individual isolators.

Two spare isolators of each type shall be stored next to installed isolators and shall be placed under the same average compressive loading as that imposed on the installed isolators. These isolators shall be retested at 10-year intervals with the same or a similar test machine and protocol used for the production tests of Section 9.5.3.

9.2.2 Isolators

9.2.2.1 Mechanical Properties. The mechanical properties of the isolation system shall not vary over the life span of the nuclear structure by more than ±20% from the values used for analysis and design, with 95% probability, accounting for variations in material properties at the time of isolator construction, aging, operating temperature, and creep.

Exception: Greater variation and/or smaller probability are acceptable provided that calculations are performed to estimate the 90th percentile displacement response of the isolation system in BDBE shaking.

9.2.2.2 Quality Assurance. A quality assurance program shall be prepared for the seismic isolators. The program shall follow ASME-NQA-1 or approved equivalent.

9.2.3 Basemat and Foundation Designs. The basemat and the foundation shall be sufficiently stiff to engage all isolators in the gravity load-resisting system and seismic force-resisting system. The basemat shall be designed to resist gravity loads, assuming the loss of one isolator caused by the local vertical settlement of the foundation below the isolation system. Multiple calculations, assuming the loss of a different isolator in the system, shall be performed.

9.3 DISPLACEMENTS AND FORCES FOR DESIGN

9.3.1 General. This section enables calculations of displacements and forces for design of the seismic isolation system (Section 9.3.2), the stop (Section 9.3.3), structures, systems, and components (SSCs) above the isolation interface (Section 9.3.4), structures below the isolation interface (Section 9.3.5), and systems and components crossing the isolation interface (Section 9.3.6). Design of systems and components below the isolation interface shall conform to other applicable chapters of this standard.

Variability in both spectral demand and isolator mechanical properties shall be considered in calculating displacements and forces for design. The ASCE 43 goals of 1% or less probability of unacceptable performance in DBE shaking and 10% or less probability of unacceptable performance in 150% DBE shaking (or BDBE shaking) shall be achieved.

Displacements and forces shall be calculated at the 80th percentile level for DBE shaking and at the 90th percentile level for BDBE shaking. DBE demands shall be calculated by response-history analysis according to Chapter 12 of ASCE 4 (2017b).

The 80th percentile displacement, D_D, and mean DBE displacement at the center of mass of the isolated superstructure, projected down to the basemat, shall be calculated and reported.

The 90th percentile BDBE displacement at the center of mass of the isolated superstructure, projected down to the basemat, D_{BD}, and the 90th percentile displacements along orthogonal horizontal axes that align with those used to define the seismic input, shall be calculated and reported. The displacements are established by either response-history analysis according to Chapter 12 of ASCE 4 (2017b) or by multiplying the mean DBE displacements by 3.0. The coexisting axial forces on the isolators shall be calculated by multiplying the mean DBE forces by the ratio, greater than 1, of the horizontal resistance of the isolation system at D_{BD} to the horizontal resistance at mean DBE displacement.

9.3.2 Seismic Isolators. The isolators shall be designed for the forces and displacements computed by response-history analysis for DBE shaking, according to Chapter 12 of ASCE 4-16 (2017b), and forces and displacements computed for BDBE shaking.

DBE displacement, D_D, shall be used for production testing of isolators. The displacement, CS, and coexisting axial gravity- and earthquake-induced forces shall be used for prototype testing of isolators.

The connections of each isolator to the superstructure and substructure shall be designed for forces associated with an isolator displacement equal to CS.

9.3.3 Stop. A stop or displacement restraint shall be provided along orthogonal horizontal axes (x, y) of an isolated superstructure to prevent excessive displacement of the isolation system in the event of BDBE shaking.

The distances between the isolated superstructure and the stop along the x- and y-axes shall be no less than the corresponding 90th percentile BDBE displacements. The lateral movement of the isolated superstructure shall not be restricted within this domain. The maximum possible displacement of the isolation system within the domain is the CS, and it shall be no less than the 90th percentile BDBE displacement.

The isolated superstructure, systems, and components do not need to be analyzed for impact loadings if CS equals or exceeds D_{BD}.

The stop shall be designed for impact loadings associated with BDBE shaking.

Exception: A smaller clearance is acceptable, provided that analysis is performed to show that impact does not result in unacceptable performance of any safety-related structures, systems, or components. Explicit analysis of the isolated superstructure, systems, and components for impact loadings is required if CS is less than D_{BD}.

9.3.4 Structures, Systems, and Components Above the Isolation Interface. The isolated superstructure, systems, and components shall be designed for the 80th percentile DBE forces and displacements computed by response-history analysis, according to Chapter 12 of ASCE 4 (2017b), using any of the analysis methods presented in Chapter 4 of this standard. Forces are permitted to be calculated according to Section 9.3.1 of this standard.

It shall be demonstrated by analysis that the performance expectations of Section 1.3 of this standard are achieved.

9.3.5 Structures Below the Isolation Interface. The foundation shall be designed for the forces delivered by the isolation system at displacement CS. Forces are permitted to be calculated according to Section 12.5.1 of ASCE 4 (2017b).

9.3.6 Systems and Components Crossing the Isolation Interface. Systems and components (or umbilical lines) that cross the isolation interface shall accommodate a displacement equal to CS and shall remain functional, with 90% or greater confidence.

9.4 PEER REVIEW

A peer review of the isolation system and the related test programs shall be performed by a team of registered design professionals. In addition to the requirements of Section 1.2.3 of ASCE 4 (2017b) and Section 9.1 of this standard, the review team shall include one or more members experienced in the application of seismic isolation and large-scale testing of isolator units.

9.5 TESTING OF PROTOTYPE AND PRODUCTION ISOLATORS

9.5.1 General. Two series of tests are described in this section: prototype and production (quality-control) tests.

9.5.2 Prototype Testing

9.5.2.1 Test Specimens. Prototype tests shall be performed separately on a minimum of three full-size specimens of each common type and size of isolator in the isolation system.

A sufficient number of prototype isolators of each common type and size shall be tested to provide 90% or greater confidence in the isolator load-carrying capacity at a lateral displacement equal to CS.

Prototype test specimens shall not be used for construction.

9.5.2.2 Record. For each cycle of each prototype test, and as a minimum, the axial force, axial displacement, horizontal force, and horizontal displacement shall be recorded digitally. The minimum sampling rate shall be 200 Hz.

9.5.2.3 Sequence and Cycles. The following minimum sequence of tests shall be performed for the prescribed number of cycles and vertical load on each prototype isolator.

1. Twenty fully reversed cycles of loading at a lateral force corresponding to the design wind force.
2. Five fully reversed cycles of loading at the design displacement, D_D.
3. Five fully reversed cycles of loading at a displacement equal to CS.

For tests 1 and 2, the imposed axial load shall equal the average dead load plus one-quarter the average live load on all isolators of that common type and size. The loading shall be (a) slow for test 1, and (b) dynamic for test 2. The frequency of loading for test 2 and bilinear isolators shall be calculated assuming the best estimate of K_d for the isolation system and the effective seismic weight of the superstructure.

For test 3, the imposed axial load shall equal either the average dead load plus one-quarter (zero) of the average live load plus (minus) the maximum expected earthquake-induced axial load or the average dead load minus the maximum expected earthquake-induced axial load for a displacement equal to CS, on all isolator units of that common type and size. The loading shall be either (a) dynamic with a frequency equal to that calculated assuming the best estimate of K_d for the isolation system and the effective seismic weight of the superstructure, or (b) slow at a frequency no greater than 0.1 Hz.

9.5.2.4 Test Specimen Adequacy. The responsible design professional shall prepare project-specific acceptance criteria for the prototype isolators. As a minimum:

1. For tests 1 and 2, the lateral force–lateral displacement relationship shall have positive stiffness on the loading segments of each cycle.
2. For test 3, the isolator unit shall sustain the imposed axial load.
3. The prototype isolator unit shall suffer no damage in tests 1 and 2. Damage is acceptable in test 3.

If any prototype isolator of a given type and size fails to meet the acceptance criteria, all isolators of that type and size shall be rejected and a minimum of another three specimens of that type and size shall be fabricated and tested.

9.5.3 Production Testing

9.5.3.1 Test Specimens. Each production isolator shall be fabricated in an identical manner and using identical materials to the corresponding prototype isolators.

Each isolator shall be subjected to a production (quality control) test.

9.5.3.2 Record. For each cycle of the production test, and as a minimum, the axial force, axial displacement, horizontal force, and horizontal displacement shall be recorded digitally. The minimum sampling rate shall be 200 Hz.

9.5.3.3 Sequence and Cycles. The following minimum sequence of tests shall be performed at a vertical load equal to the average dead load plus one-quarter the effects caused by live load on all isolators of a common type and size:

1. Five fully reversed cycles of loading at displacement D_d.

The loading shall be either (a) dynamic, with a frequency equal to that calculated for the isolation system assuming the best-estimate second-slope stiffness for the isolation system and the effective seismic weight of the superstructure, or (b) slow, at a frequency no greater than 0.1 Hz. If slow-speed tests are performed, the mechanical properties at slow speed shall be correlated to the dynamic properties using full-scale test data.

9.5.3.4 Test Specimen Adequacy. The responsible design professional shall prepare project-specific acceptance criteria for the production isolator units. As a minimum:

1. For test 1, the lateral force–lateral displacement relationship shall have positive stiffness on the loading segments of each cycle.
2. The production isolator unit shall suffer no damage in test 1.

CHAPTER 10
QUALITY ASSURANCE PROVISIONS

The seismic analysis and design of nuclear facilities specified in this standard shall be performed under a documented quality assurance (QA) program. Verification of both design and construction shall be performed to ensure acceptable seismic behavior and shall be an integral part of the QA process.

10.1 DESIGN VERIFICATION AND INDEPENDENT PEER REVIEW

10.1.1 Seismic Design Verification. Individuals or groups other than those who performed the design work shall perform design verification. Design verification shall include provisions for verifying and checking the adequacy of the analysis and design by any one, or a combination of, the following:

- Using alternate or simplified calculation methods,
- Using a suitable testing program, or
- Reviewing the original design and verifying the underlying assumptions, methods, and inputs to be correct and outputs to be reasonable.

10.1.2 Independent Seismic Peer Review. An independent seismic peer review shall be conducted for seismic design category (SDC) SDC 3, SDC 4, and SDC 5 structures, systems, and components (SSCs). For an SDC 2 SSC, the design organization shall determine if a graded seismic peer review is necessary based on the importance of the SSC to hazard mitigation and design complexity. Independent seismic peer review is a form of design review and shall be conducted by qualified individuals with knowledge of the unique features of the design and analysis, as well as the type of construction used. The peer review program shall be developed using a graded approach, paralleling the seismic design categories of SDC 3, SDC 4, and SDC 5 so that the level of review is commensurate with the complexity of the design, the number of design disciplines involved, and any nontraditional analysis and design concepts used. The peer reviewers shall be independent from the design verifiers.

10.2 STRUCTURAL OBSERVATION, INSPECTION, AND TESTING

Nuclear facilities shall be constructed in accordance with the design documents to ensure that SSCs respond to an earthquake as intended. A construction verification program shall be implemented that is consistent with the applicable provisions of the selected design code and that is based on a graded approach that addresses the elements described in the following subsections.

10.2.1 Structural Observations. The Engineer of Record (EOR) or his or her designee(s) shall perform structural observations of the construction processes involving observations of the structural system at significant construction stages for conformance to the approved plans and specifications. The stages of structural observations, their scope, and the type of records to be kept shall be identified in the construction verfication program. Records of the observation and resolution of any comments shall be entered into the project records.

10.2.2 Continuous and Periodic Inspections. Qualifications of personnel performing inspections shall meet the requirements of the code specified by the design organization. Duties include performing visual inspections and field measurements of materials, obtaining specimens for testing, and preparing reports. The construction verification program shall identify the items requiring both continuous and periodic inspections. It shall also include a schedule of inspections and shall specify the records to be kept and the reports to be filed. The inspection program shall include foundation and backfill soils, reinforced concrete, structural masonry, and steel structures, as well as anchorage and supports that are part of these structures. It shall also identify permissible equipment to be used and standards to be followed for processes subject to inspection. Items requiring continuous and periodic examination shall be clearly identified, including the frequency of inspection.

10.2.3 Testing. The construction verification program shall identify in-progress testing requirements and methods of testing. All testing shall be performed by testing laboratories under the responsible charge of a licensed professional engineer, approved by the design organization. Testing shall include measuring, examining, testing, calibrating, or otherwise determining the characteristics or performance of construction materials and verifying compliance with construction documents. The construction verification program shall also specify the number and frequency of test specimens taken in accordance with the applicable codes and standards or as specified in the approved plans, whichever is more stringent. Records of inspection and testing shall be reviewed by the design organization and shall be made a part of the permanent records for the facility.

10.3 QUALITY ASSURANCE PROGRAM

A QA program that, at a minimum, includes requirements for organization, training and qualifications, work processes, documents and records (e.g., specifications, drawings, procedures, and instructions), and assessments shall be implemented by projects designed in accordance with this standard. The QA program shall include the minimum required elements described in the following subsections.

A graded approach shall be used in establishing the QA program with increasing rigor for SDC 2 to SDC 5 SSCs.

All calculations, including computer-generated ones. shall be documented in sufficient detail for a reviewer to determine that the design requirements have been correctly identified and implemented. A graded approach shall be considered for the level of detail and rigor in the documentation.

10.3.1 Design Basis Documents. Design basis documents shall provide details on the following:

- Design organization, including design team, division of responsibility, team interface control, and organizational procedures;
- Standards, instructions, procedures, and drawings;
- Design input and specification (criteria documents), including design basis loads, interfaces, load combinations, and acceptable structural performance (i.e., acceptable ductility for structures and structural subsystems);
- Industry standards and codes applicable to the design;
- Design assumptions and methodology for structures and structural subsystems;
- Design report to the owner; and
- Design output documents.

10.3.2 Design Procedures. Procedures are required that provide for the following:

- Verification and validation (V&V) of analytical software,
- Control and updating of software,
- Design approval and reports,
- Design change and document control,
- Design audits and/or independent assessments, and
- Design verification methods.

10.4 SOFTWARE QUALITY ASSURANCE

All computer programs used in the analysis and design of SSCs in nuclear facilities shall be controlled, verified, and validated before use, or the computer program results shall be independently verified and validated for each application. The V&V shall ensure that

1. The computer program produces correct solutions for the applied mathematical model within defined limits for parameters of interest, and
2. The mathematical model produces valid solutions for the design components and phenomena being modeled.

A software V&V plan shall be developed that

1. Identifies significant features of the program that need to be verified for the project,
2. Identifies the types of test or benchmark problems that are needed to implement the plan,
3. Establishes the acceptance limits within which computed results could vary from the closed-form results or other verified results,
4. Includes for validation purposes a full suite of test cases covering all selected program features used, and
5. Provides sample problems with verified solutions against which the computer program installation can be validated for each computer on which the program is used.

In addition, computer programs used in SDC 3, SDC 4, and SDC 5 SSCs that are not developed or not acquired under a QA program approved for the project shall be subjected to an approved dedication activity. Alternatively, in lieu of dedication, the results of such computer programs shall be independently verified and validated for each application.

APPENDIX A

ALTERNATE METHOD TO MEET ASCE 43 PERFORMANCE GOALS WHEN SEISMIC CAPABILITIES ARE DEFINED AT THE 50% PROBABILITY OF FAILURE LEVEL

A1. INTRODUCTION

For some seismically induced failure modes (primarily soil failure modes), it is common engineering practice to define the seismic capability at the 50% probability of failure level (called herein *median capacity*, $C_{50\%}$) and to apply an appropriate factor of safety. This appendix defines an acceptable method to meet the ASCE 43 performance goals, P_F, when the seismic capability is defined at the median capacity level, $C_{50\%}$.

A2. SPECIFICATION OF THE DESIGN RESPONSE SPECTRUM

The design response spectrum (DRS) DRS_A, for use with median capacities is defined by

$$\mathrm{DRS}_A = F_{50\%} \times \mathrm{UHRS}_{H_P} \quad \text{(A-1)}$$

where UHRS_{H_P} is the uniform hazard response spectrum defined at the mean annual frequency of exceedance $H_P = P_F$.

A3. SPECIFICATION OF MINIMUM FACTOR OF SAFETY TO BE APPLIED TO MEDIAN PROBABILITY OF FAILURE CAPABILITY

The minimum required factor of safety, $F_{50\%}$, to be applied to the median capacity, $C_{50\%}$, is defined by

$$F_{50\%} = \exp\left[\frac{1}{2} K_H \beta^2\right] \quad \text{(A-2)}$$

where β defines the natural logarithmic standard deviation of the capacity, and

$$K_H = \frac{1}{\log A_R} \quad \text{(A-3)}$$

where A_R is defined in Section 2.2 of the standard. Because $F_{50\%}$ is particularly sensitive to β, either a realistic or conservative estimate of β shall be made.

APPENDIX B

ALTERNATE METHOD TO MEET ASCE 43 PERFORMANCE GOALS WHEN SEISMIC CAPABILITIES ARE DEFINED AT THE 10% PROBABILITY OF FAILURE LEVEL

B1. INTRODUCTION

For some seismically induced failure modes (primarily soil failure modes), it can be engineering practice to define the seismic capability at the 10% probability of failure level $C_{10\%}$ and to apply an appropriate factor of safety. This appendix defines an acceptable method to meet the ASCE 43 performance goals, P_F, when the seismic capability is defined at the 10% probability of failure level, $C_{10\%}$.

B2. SPECIFICATION OF THE DESIGN RESPONSE SPECTRUM

The design response spectrum (DRS), DRS_B, for use with capabilities defined at the 10% probability of failure level, $C_{10\%}$, is given by

$$DRS_B = F_{10\%} \times DRS \quad \text{(B-1)}$$

The DRS is defined in Section 2.2 of this standard in Equations (2-1) through (2-3).

B3. SPECIFICATION OF MINIMUM FACTOR OF SAFETY TO BE APPLIED TO 10% PROBABILITY OF FAILURE CAPABILITY

The minimum required factor of safety, $F_{10\%}$, to be applied to the 10% probability of failure capability, $C_{10\%}$, is defined by

$$F_{10\%} = 1.50 \quad \text{(B-2)}$$

This minimum factor of safety requirement can be achieved by increasing the DRS by the factor of safety, $F_{10\%}$.

REFERENCES

AASHTO. 2010. *Guide specification for seismic isolation design*. Washington, DC: AASHTO.

ACI (American Concrete Institute). 2007. *Qualification of post-installed mechanical anchors in concrete and commentary*. ACI 355.2-07. Farmington Hills, MI: ACI.

ACI. 2008. *Building code requirements for structure concrete and commentary*. ACI 318-08. Farmington Hills, MI: ACI.

ACI. 2011. *Qualification of post-installed adhesive anchors in concrete and commentary*. ACI 355.4-11. Farmington Hills, MI: ACI.

ACI. 2013a. *Building code requirements and specification for masonry structures and companion commentaries*. ACI 530/530.1-13. Farmington Hills, MI: ACI.

ACI. 2013b. *Code requirements for nuclear safety-related concrete structures and commentary*. ACI 349-13. Farmington Hills, MI: ACI.

ACI. 2014. *Building code requirements for structural concrete and commentary*. ACI 318-14. Farmington Hills, MI: ACI.

ACI. 2015. *Code for concrete containments*. ACI 359-15. Farmington Hills, MI: ACI.

AISC (American Institute of Steel Construction). 2016a. *Specification for structural steel buildings*. ANSI/AISC 360. Chicago: AISC.

AISC. 2011. *Steel construction manual*. ANSI/AISC 325-11. Chicago: AISC.

AISC. 2012. *Specification for safety-related steel structures for nuclear facilities*. ANSI/AISC N690-12. Chicago: AISC.

AISC. 2013. *Manual of steel construction*, 14th ed. Chicago: AISC.

AISC. 2015. *Specification for safety-related steel structures for nuclear facilities, January 31, 2012, incl. Supplement No. 1, August 11, 2015*. ANSI/AISC N690-12 and ANSI/AISC N690s1-15. Chicago: AISC.

AISC. 2016b. *Prequalified connections for special and intermediate steel moment frames for seismic applications*. ANSI/AISC 358-16. Chicago: AISC.

AISC. 2016c. *Seismic provisions for structural steel buildings*. ANSI/AISC 341-16. Chicago: AISC.

AISC. 2018. *Specification for safety-related steel structures for nuclear facilities*. ANSI/AISC N690-18. Chicago: AISC.

AISI (American Iron and Steel Institute). 2012. *North American specification for the design of cold-formed steel structural members*. AISI S100-12. Washington, DC: AISC.

Akiyama, H., Sekimoto, H., Tanaka, M., Inoue, K., Fukihara, M., and Okuda, Y. 1989. "1/10th Scale model test of inner concrete structure composed of concrete filled steel bearing wall." In *Proc., 10th Int. Conf. on Structural Mechanics in Reactor Technology* (SMiRT 10). Anaheim, California.

ANS (American Nuclear Society). 2016a. *Criteria for investigations of nuclear facility sites for seismic hazard assessments*. ANSI/ANS 2.27-2008. La Grange Park, IL: ANS.

ANS. 2016b. *Probabilistic seismic hazard analysis*. ANSI/ANS 2.29-2008. La Grange Park, IL: ANS.

ANS. R2016. *Categorization of nuclear facility structures, systems, and components for seismic design*. ANSI/ANS-2.26-2004. La Grange Park, IL: ANS.

ASCE. R2017. *Guideline for design and analysis of nuclear safety-related earth structures*. ANSI/ASCE 1-82. Reston, VA: ASCE.

ASCE. 2000. *Quality in the constructed project: A guide for owners, designers, and constructors: ASCE Manuals and reports on engineering practice no. 73*. Reston, VA: ASCE.

ASCE. 2002. *Specification for the design of cold-formed stainless steel structural members*. ASCE/SEI 8-02. Reston, VA: ASCE.

ASCE. 2005. *Seismic design criteria for structures, systems, and components in nuclear facilities*. ASCE/SEI 43-05. Reston, VA: ASCE.

ASCE. 2013. *Seismic evaluation and retrofit of existing buildings*. ASCE/SEI 41-13. Reston, VA: ASCE.

ASCE. 2017a. *Minimum design loads and associated criteria for buildings and other structures*. ASCE/SEI 7-16. Reston, VA: ASCE.

ASCE. 2017b. *Seismic analysis of safety-related nuclear structures*. ASCE/SEI 4-16. Reston, VA: ASCE.

ASME (American Society of Mechanical Engineers.) 2017a. *Qualification of active mechanical equipment used in nuclear power facilities*. ASME QME-1. New York: ASME.

ASME. 2017b. *Quality assurance requirements for nuclear facility applications*. ASME NQA-1. New York: ASME.

ASME. 2019a. *Boiler and pressure vessel code (BPVC)*. New York: ASME.

ASME. 2019b. "Rules for evaluation of service loadings with Level D service limits." In *Boiler and pressure vessel code*. New York: ASME.

ASME. 2019c. "Seismic load coefficient method for piping systems." In *Boiler and pressure vessel code*. New York: ASME.

ATC (Applied Technology Council). 1974. *An evaluation of response spectrum approach to seismic design of buildings*. ATC-2. Redwood City, CA: ATC.

Barda, F., Hanson, J. M., and Corley, W. G. 1977. "Shear strength of low-rise walls with boundary elements." ACI Special Publication 53: 149–202.

Bozard, E. S. 2002. *The influence of soil-structure-interaction on the inelastic force reduction factor, F_{μ}*. WSRC-TR-2002-00333. Aiken, SC: Savannah River Site.

Constantinou, M. C., Tsopelas, P., Kasalanati, A., and Wolff, E. D. 1999. *Property modification factors for seismic isolation bearings*. Technical Rep. No. MCEER-99-0012. Buffalo, NY: State Univ. of New York.

Constantinou, M. C., Whittaker, A. S., Kalpakidis, Y., Fenz, D. M., and Warn, G. P. 2007. *Performance of seismic isolation hardware under service and seismic loading*. Technical Rep. No. MCEER-07-0012. Buffalo, NY: State Univ. of New York.

Costantino, C. J., and Houston, T. W. 2008. *Summary of calculations made in support of modifications to standard review plan sections 3.7.1, 3.7.2 and 3.7.3*. Spring Valley, NY: C. J. Costantino and Associates.

Duffey, T. A., Goldman, A., and Farrar, C. R. 1994. *Shear wall ultimate drift limits*. Rep. No. NUREG/CR-6104. Rockville, MD: US Nuclear Regulatory Commission.

Duncan, J. M., and Seed, R. B. 1986. "Compaction induced earth pressures under K0 conditions." *J. Geotech. Eng.* 112 (1): 1–22.

Duncan, J. M., Williams, G. W., Sehn, A. L., and Seed, R. B. 1991. "Estimation of earth pressures due to compaction." *J. Geotech. Eng.* 117 (12): 1833–1847.

Eggers, D., and Baig, M. 2015. "Significance of ratcheting in seismic analyses." In *Proc., 23rd Int. Conf. on Structural Mechanics in Reactor Technology (SMiRT 23)* Manchester, UK.

Epackachi, S., Nguyen, N. H., Kurt, E. G., Whittaker, A. S., and Varma, A. H. 2015a. "In-plane seismic behavior of rectangular steel-plate composite wall piers." *J. Struct. Eng.* 141 (7).

Epackachi, S., Whittaker, A. S., and Huang, Y. N. 2015b. "Analytical modeling of rectangular SC wall panels." *J. Constr. Steel Res.* 105: 49–59.

Epackachi, S., Whittaker, A. S., Varma, A. H., and Kurt, E. G. 2015c. "Finite element modeling of steel-plate concrete composite wall piers." *Eng. Struct.* 100 (Oct): 369–384.

EPRI (Electric Power Research Institute). 1991a. *Generic equipment ruggedness of power plant equipment (Revision 1)*. Rep. No. NP-5223-SLR1. Palo Alto, CA: EPRI.

EPRI. 1991b. *Summary of the seismic adequacy of twenty classes of equipment required for the safe shutdown of nuclear plants*. Rep. No. EPRI NP-7149-D. Palo Alto, CA: EPRI.

EPRI. 1996. *GERS formulated using data from the SQURTS program*. Rep. No. TR-105988. Palo Alto, CA: EPRI.

FEMA. 1997. *NEHRP guidelines for the seismic rehabilitation of buildings*. FEMA 273. Washington, DC: Applied Technology Council for Building Seismic Safety Council.

FEMA. 2009. *Quantification of building seismic performance factors*. FEMA P695. Washington, DC: Applied Technology Council.

Gulec, C. K., Whittaker, A. S., and Stojadinovic, B. 2008. "Shear strength of squat rectangular reinforced concrete walls." *ACI Struct. J.* 105 (4): 488–497.

Gulec, C. K., Whittaker, A. S., and Stojadinovic, B. 2009. "Peak shear strength of squat reinforced concrete walls with boundary barbells or flanges." *ACI Struct. J.* 106 (3): 368–377.

Hadjian, A. H. 1981. "On the correlation of the components of strong ground motion—Part 2." *Bull. Seismol. Soc. Am.* 71 (4): 1323–1331.

Huang, Y.-N., Whittaker, A. S., Kennedy, R. P., and Mayes, R. L. 2009. *Assessment of base-isolated nuclear structures for design and beyond-design basis earthquake shaking*. Rep. No. MCEER-09-0008. Buffalo, NY: State Univ. of New York.

Huang, Y.-N., Whittaker, A. S., Kennedy, R. P., and Mayes, R. L. 2012. "Response of base-isolated nuclear structures for design basis and beyond design basis earthquake shaking." *Earthquake Eng. Struct. Dyn.* 42 (3): 339–356.

Huang, Y.-N., Whittaker, A. S., and Luco, N. 2008. *Performance assessment of conventional and base-isolated nuclear power plants for earthquake and blast loadings*. Rep. No. MCEER-08-0019. Buffalo, NY: State Univ. of New York.

Huang, Y.-N., Whittaker, A. S., and Luco, N. 2010. "Seismic performance assessment of base-isolated safety-related nuclear structures." *Earthquake Eng. Struct. Dyn.* 39 (13): 1421–1442.

Huang, Y.-N., Whittaker, A. S., and Luco, N. 2011. *Establishing maximum spectral demand for performance-based earthquake engineering: Collaborative research with the University of Buffalo and the USGS*. Technical Rep. Award No. 08HQGR0017. Reston, VA: USGS.

Huang, Y.-N., Yen, W.-Y., and Whittaker, A. S. 2016. "Correlation of horizontal and vertical components of strong ground motion for response-history analysis of safety-related nuclear structures." *Nucl. Eng. Des.* 310 (Dec): 273–279.

ICC (International Code Council). 2015. *International Building Code*. Whittier, CA: ICC.

IEEE. 2006. *IEEE standard for qualification of safety-related actuators for nuclear power generating stations*. IEEE 382-2006. New York: IEEE.

IEEE. 2013. *IEEE standard for seismic qualification of equipment for nuclear power generating stations*. IEEE 344-2013. New York: IEEE.

Kalpakidis, I. V., Constantinou, M. C., and Whittaker, A. S. 2010. "Modeling strength degradation in lead-rubber bearings under earthquake shaking." *Earthquake Eng. Struct. Dyn.* 39 (13): 1533–1549.

Kennedy, R. P. 2011. "Performance-goal based (risk informed) approach for establishing the SSE site specific response spectrum for future nuclear power plants." *Nucl. Eng. Des.* 241 (3): 648–656.

Kennedy, R. P., Short, S. A., Merz, K. L., Tokarz, F. J., Idriss, I. M., Power, M. S., et al. 1984. *Engineering characterization of ground motion: Task 1: Effects of characteristics of free-field motion on structural response*. Rep. No. NUREG/CR-3805. Rockville, MD: US Nuclear Regulatory Commission.

Kennedy, R. P., Wesley, D. A., and Tong, W. H. 1988. *Probabilistic evaluation of the Diablo Canyon turbine seismic capacity using nonlinear time history analyses*. Rep. No. 1643.01. San Francisco: Pacific Gas and Electric Company.

Kumar, M., Whittaker, A. S., and Constantinou, M. C. 2014. "An advanced numerical model of elastomeric seismic isolation bearings." *Earthquake Eng. Struct. Dyn.* 43 (13): 1955–1974.

Kumar, M., Whittaker, A. S., and Constantinou, M. C. 2015a. *Seismic isolation of safety-related nuclear facilities using sliding bearings*. Rep. No. MCEER-15-0006. Buffalo, NY: State Univ. of New York.

Kumar, M., Whittaker, A. S., and Constantinou, M. C. 2015b. *Seismic isolation of safety-related nuclear facilities using elastomeric bearings*. Rep. No. MCEER-15-0008. Buffalo, NY: State Univ. of New York.

McGuire, R. K., Silva, W. J., and Costantino, C. J. 2001. *Technical basis for revision of regulatory guidance on design ground motions: Hazard- and risk-consistent ground motion spectra guidelines*. Rep. No. NUREG/CR-6728. Rockville, MD: US Nuclear Regulatory Commission.

McGuire, R. K., Silva, W. J., and Costantino, C. J. 2002. *Technical basis for revision of regulatory guidance on design ground motions: Hazard- and risk-consistent ground motion spectra for two sites*. Rep. No. NUREG/CR-6769. Rockville, MD: US Nuclear Regulatory Commission.

Mertz, G. E. 2001. *Force reduction factors for the structural design and evaluation of facilities containing nuclear and hazardous materials*. Rep. No. WSRC-TR-2001-00037. Aiken, SC: Savannah River Site.

Mertz, G. E., Spears, R., and Houston, T. W. 2016. "The effects of discretization errors on the high frequency content of in-structure-response spectra." In *Proc., 2016 Pressure Vessel and Piping Conf.* Vancouver.

Miranda, E., and Bertero, V. V. 1994. "Evaluation of strength reduction factors for earthquake-resistant design." *Earthquake Spectra* 10 (2): 357–379.

Moehle, J. P., Hooper, J. D., Kelly, D. J., and Meyer, T. R. 2010. *Seismic design of cast-in-place concrete diaphragms, chords, and collectors: A guide for practicing engineers.* NEHRP Seismic Design Technical Brief No. 3, NIST GCR 10-917-4. Gaithersburg, MD: NIST.

Morante, R. J. 2006. *Recommendations for revision of seismic damping values in Regulatory Guide 1.61.* Rep. No. NUREG/CR-6919. Rockville, MD: USNRC.

Morgan, T., Whittaker, A. S., and Thompson, A. C. 2001. "Cyclic behavior of high-damping rubber bearings." In *Proc., 5th World Congress on Joints, Bearings and Seismic Systems for Concrete Structures*, Rome.

Naeim, F., and Kelly, J. M. 1999. *Design of seismic isolated structures: From theory to practice.* New York: Wiley.

Nassar, A. A., and Krawinkler, H. 1991. *Seismic demands for SDOF and MDOF systems.* Rep. No. 95. Redwood City, CA: Stanford Univ.

Newmark, N. M., and Hall, W. J. 1978. *Development of criteria for seismic review of selected nuclear power plants.* Rockville, MD: US Nuclear Regulatory Commission.

Newmark, N. M., and Hall, W. J. 1987. *Earthquake spectra and design.* Oakland, CA: Earthquake Engineering Research Institute.

Ostadan, F., and White, W. H. 1998. "Lateral seismic soil pressure: An updated approach." In *Proc., US–Japan Soil Structure Interaction Workshop.* Menlo Park, California.

Osteraas, J. D., and Krawinkler, H. 1990. *Strength and ductility considerations in seismic demand.* Rep. No. 90. Redwood City, CA: Stanford Univ.

Reineck, K.-H., Bentz, E. C., Fitik, B., Kuchma, D. A., and Bayrak, O. 2013. "ACI-DAfStb database of shear tests on slender reinforced concrete beams without stirrups." *ACI Struct. J.* 110 (5): 867–876.

Riddell, R., and Newmark, N. M. 1979. *Statistical analysis of the response of nonlinear systems subjected to earthquakes: Structural research series No. 468.* Urbana, IL: Univ. of Illinois, Dept. of Civil Engineering.

Salmon, M. W., and Kennedy, R. P. 1994. *Meeting performance goals by use of experience data.* UCRL-CR-120813. Livermore, CA: Lawrence Livermore National Laboratory.

Seneviratna, G. D. P. K., and Krawinkler, H. 1993. *Effects of soft soil and hysteresis model on seismic demand.* Rep. No. 108. Redwood City, CA: Stanford Univ.

Seneviratna, G. D. P. K., and Krawinkler, H. 1997. *Evaluation of inelastic MDOF effects for seismic demand.* Rep. No. 120. Redwood City, CA: Stanford Univ.

Seo, J., Varma, A. H., Sener, K., and Ayhan, D. 2016. "Steel plate composite (SC) walls: In-plane shear behavior, database, and design." *J. Construct. Steel Res.* 119 (2016): 202–215.

SQUG (Seismic Qualification Utility Group). 2000. *Implementation guidelines for seismic qualification of new and replacement equipment/parts (NARE) using the generic implementation procedure (GIP).* Rev. 4. Palo Alto, CA: Electric Power Research Institute.

SQUG. 2001. *Generic implementation procedure (GIP) for seismic verification of nuclear plant equipment.* Rev. 3a. Palo Alto, CA: Electric Power Research Institute.

Terzaghi, K., Peck, R. B., and Mesri, G. 1995. *Soil mechanics in engineering practice*, 3rd ed. New York: Wiley.

Thompson, A., Whittaker, A. S., Fenves, G. L., and Mahin, S. A. 2000. "Property modification factors for elastomeric bearings." In *Proc., 12th World Conf. on Earthquake Engineering*, Auckland, NZ.

TMS (The Masonry Society). 2016. *Building code requirements and specifications for masonry structures.* TMS 402/602. Longmont, CO: TMS.

US Code of Federal Regulations. 2011. *Quality assurance requirements.* 10 CFR 830, Subpart A. Washington, DC: Government Printing Office.

USDOE (US Department of Energy). 1997. *Seismic evaluation procedure for equipment in US Dept. of Energy facilities.* DOE/EH-0545. Washington, DC: USDOE.

USDOE. 2013. *Quality assurance program guide.* DOE G 414.1-2B. Washington, DC: USDOE.

USNRC (US Nuclear Regulatory Commission). 1992. *NRC supplemental safety evaluation reports on the GIP.* SSER No. 2, Rev. 2 and SSER No. 3, Rev. 3. Rockville, MD: USNRC.

USNRC. 1997a. *Identification and characterization of seismic sources and determination of safe shutdown earthquake ground motion.* Regulatory Guide 1.165. Rockville, MD: USNRC.

USNRC. 1997b. NRC Supplemental safety evaluation reports on the AIP SEER No. 3 Rockville, Maryland: USNRC.

USNRC. 2008. *Quality assurance criteria for nuclear power plants and fuel reprocessing plants.* 10 CFR 50, Appendix B. Rockville, MD: USNRC.

USNRC. 2014a. *Standard review plan section 3.7.1.* NUREG-0800, Rev. 4. Rockville, MD: USNRC.

USNRC. 2014b. *Technical rationale for enhancements to seismic and structural review guidance.* Rev. 1. Rockville, MD: USNRC.

Varma, A. H., Malushte, S. R., Sener, K., and Lai, Z. 2013. "Steel-plate composite (SC) walls for safety related nuclear facilities: Design for in-plane force and out-of-plane moments." *Nucl. Eng. Des.* 269 (1): 240–249.

Varma, A. H., Zhang, K., Chi, H., Booth, P., and Baker, T. 2011. "In-plane shear behavior of SC composite walls: Theory vs. experiment." In *Proc., 21st Int. Conf. on Structural Mechanics in Reactor Technology (SMiRT 21).* New Delhi.

Warn, G. P., and Whittaker, A. S. 2004. "Performance estimates in seismically isolated bridges." *Eng. Struct.* 26 (9): 1261–1278.

Whittaker, A. S., Constantinou, M., and Tsopelas, P. 1998. "Displacement estimates for performance-based seismic design." *J. Struct. Eng.* 124 (8): 905–912.

Yim, S. C.-S., and Chopra, A. K. 1985. "Simplified earthquake analysis of multistory structures with foundation uplift." *J. Struct. Eng.* 111 (12): 2708–2731.

CHAPTER C1
INTRODUCTION

C1.3 SEISMIC DESIGN CRITERIA

The objective of this standard is to provide design criteria, which when properly implemented in design and construction, produce seismically robust structures, systems, and components (SSCs). Seismically robust structures, systems, and components safeguard against the accidental release of nuclear material to the workers, the public, and the environment. The criteria focus on cast-in-place reinforced concrete and structural steel systems. The construction details required for these structural systems, according to referenced materials standards such as ACI 349 (ACI 2013b) and AISC N690 (AISC 2018), should ensure seismic ruggedness with a high degree of reliability. This design standard should be read in conjunction with ASCE 1 (ASCE 1982), ASCE 4 (ASCE 2016b), and other codes and standards referenced herein.

Requirements for structural systems with a demonstrated poor seismic performance record, such as unreinforced masonry or precast concrete using conventional gravity connections, have been purposely omitted from this standard and should not be used in the primary seismic load path.

The criteria presented in this standard do not discourage the use of any structural system with reliable seismic performance. Guidance is provided (in Section 1.5 and in the commentary) to develop criteria for seismically rugged structural systems not discussed herein that ensure the same degree of safety as those structural systems explicitly addressed.

This standard prescribes design criteria that are graded according to tolerable risk. Those SSCs whose failure results in more serious undesirable consequences to the facility workers, the public, or the environment are designed for more stringent criteria, as measured by seismic design category (SDC) and limit state (LS), as defined in ANS 2.26 (ANS 2017) and as described in this commentary. Four of the five SDCs identified in ANS 2.26 are considered in this standard: SDC 2, SDC 3, SDC 4, and SDC 5, as listed in Table 1-1. All four of the limit states identified in ANS 2.26 are considered in this standard, LS A, LS B, LS C, and LS D, as listed in Table 1-2.

The seismic performance of an individual SSC is defined by two measures that together form a seismic design basis (SDB): a qualitative description of the acceptable level of damage, and a quantitative annual frequency that damage will exceed the acceptable level.

The qualitative description of acceptable damage is referred to as an LS. The four LSs are reproduced from ANS 2.26 in Table 1-2 and range from A (significant damage) to D (negligible damage). Design loads in Chapters 5 and 8 and deformation limits in Chapter 5 are used to limit SSC damage to the degree associated with the assigned LS. Components, equipment, and distribution systems assigned lower tolerable damage are designed for higher seismic loads. Structural systems with lower tolerable damage should satisfy more stringent (smaller) criteria for drift ratios (Table 5-2) and plastic hinge rotations (Table 5-3) than those assigned higher tolerable damage.

For each SSC, the maximum annual frequency of seismic damage exceeding the limit state is the target performance goal (P_F). Table 1-1 presents performance goals for SDC 2 (4×10^{-4}) through SDC 5 (1×10^{-5}). Design basis earthquake (DBE) shaking is defined for each SDC. Chapter 2 presents criteria to define DBE shaking. The frequency of exceeding an LS is a function of fragility, seismic hazard, and the rate of change of the hazard. Limit state exceedance frequency is calculated by integrating a fragility curve over the derivative of a seismic hazard curve, where the two curves share the same abscissa (e.g., spectral acceleration at 5 Hz). Rigorous modern probabilistic seismic hazard results, including the maps prepared by the US Geological Survey (USGS), indicate that the rate of change of ground motion intensity versus exceedance frequency is not constant throughout the United States. As a consequence, design to a prescribed ground motion hazard would result in varying LS exceedance probabilities at different locations for SSCs with an identical fragility. Figure C1-1 plots the 2014 USGS 5 Hz spectral acceleration versus the annual frequency of exceedance for sites with a near-surface shear wave velocity of 760 m/s (2,490 ft/s). The curves have been normalized at 2% probability of exceedance in 50 years. Figure C1-1 shows that the slope of the hazard curve, and therefore the frequency of exceedance of an LS, varies across the country. For instance, in coastal California (e.g., at SLAC National Accelerator Laboratory in Menlo Park), the ratio of the spectral acceleration associated with a 10-fold increase in hazard ranges from about 1.5 to 3.5, whereas in the Central and Eastern United States (e.g., Oak Ridge, Tennessee) those ratios range from 2.5 to 5.

Seismic performance can be improved by decreasing the level of allowable damage (LS), by decreasing the annual frequency of exceeding that damage (P_F), or by some combination of the two. For instance, a structure to be designed to have an annual probability of 4×10^{-5} (SDC 4) of suffering damage that exceeds limited permanent distortion (LS C) would be assigned SDB 4C. The expected performance could be improved by changing the LS to D (the structure remains essentially elastic) or by changing the exceedance probability to 1×10^{-5} (SDC 5). In this standard, the most stringent seismic requirements are associated with SDB-5D, and the least stringent are associated with SDB-2A.

The performance goals of Table 1-1 represent the annual frequency of unacceptable performance (as defined by its LS) for each safety-related SSC. As such, these SSC-specific performance goals cannot be compared with performance goals (e.g., collapse prevention or life safety) for commercial (nonnuclear) buildings. Importantly, the performance goal of 4×10^{-4} for an

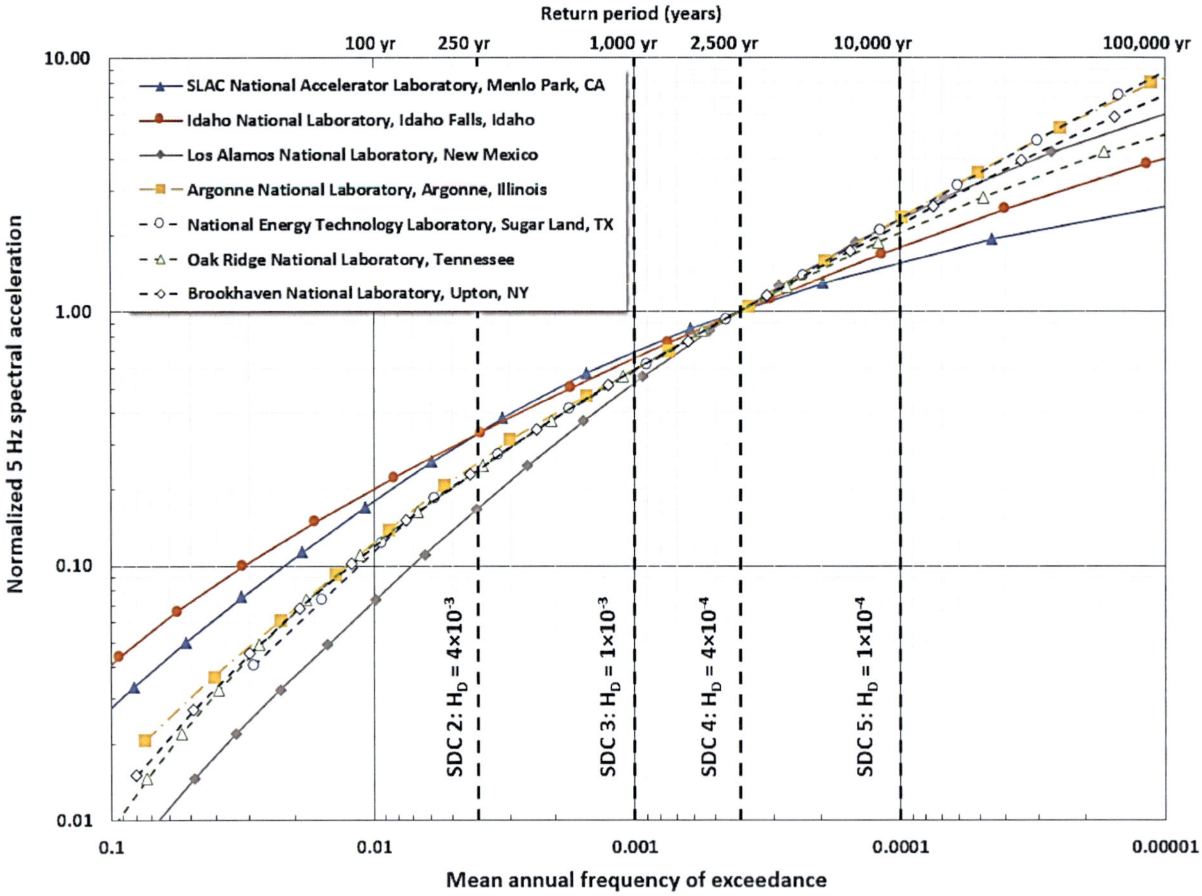

Figure C1-1. Normalized 5 Hz spectral acceleration at selected US DOE sites.

SDC 2 SSC in this standard cannot be compared to the building-level collapse prevention goal of ASCE 7.

Chapter 2 of this standard provides a method to define DBE response spectra. Load effects on SSCs are calculated by analysis of a building model according to ASCE 4 for the seismic demands of Chapter 2, as modified by inelastic energy absorption factors, according to Chapters 5 and 8. Detailing according to the requirements of the referenced materials standards, as modified by Chapter 6, should enable an SSC to achieve its seismic design basis.

C1.4 INTEGRATION OF OTHER CODES AND STANDARDS WITH ASCE 43

Specific ANS standards that address seismic design of SSCs in nuclear safety-related facilities are the following:

- ANSI/ANS 2.26-R2017 (ANS 2017), *Categorization of Nuclear Facility Structures, Systems, and Components for Seismic Design* provides guidelines and criteria for selecting SDCs and limit states, which are inputs to this standard. ANSI/ANS 2.26 identifies five SDCs (SDC 1 to SDC 5) based on the importance of the consequences of failure and defines four limit states (A to D) based on damage and ability to perform safety function(s). ANSI/ANS 2.26 uses a graded approach to ensure that the level of conservatism and rigor in design is appropriate for facility characteristics, such as hazards to workers, the public, and the environment. The combination of SDC and limit state defines the seismic design basis (SDB) for each SSC. Thus, an SSC with SDB-3C would use criteria for SDC 3 and Limit State C. A total of 20 SDBs are defined in ANS 2.26 that can match seismic design criteria to SSC safety function and importance, implementing a graded approach.

  The seismic design criteria and methods specified in this standard (ASCE 43), however, are intended to achieve the same performance goals given in ANS 2.26-R2017 for SDC 3 through SDC 5. Both ANS 2.26 and the previous version of this standard defer seismic design of SDC 1 and SDC 2 SSCs to the International Building Code, which in turn points to ASCE 7 (ASCE 2016a). This standard now includes design provisions for SDC 2 SSCs using a newly introduced performance goal and the associated seismic design criteria. This standard (ASCE 43) does not discuss or provide seismic design criteria for SDC 1 SSCs.
- ANSI/ANS 2.27-R2016 (ANS 2016a), *Criteria for Investigations of Nuclear Facility Sites for Seismic Hazard Assessments* provides requirements and recommended practices for conducting site-specific investigations and acquiring data sets needed to characterize seismic sources for probabilistic seismic hazard analysis. The data sets provide information for site response and soil–structure interaction analyses needed for design of nuclear facilities. It requires site investigations "in sufficient scope and detail necessary to support the evaluations required by ANSI/ANS-2.29 and ASCE 43-05 and to support the objectives of ANSI/ANS-2.26." Accordingly, it provides guidance for selecting different levels of site investigation based on the highest SDC of SSCs at the site and using a graded approach.

- ANSI/ANS 2.29-R2016, *Probabilistic Seismic Hazard Analysis* (ANS 2016b) establishes requirements for performing these calculations. Like ANSI/ANS 2.27, it requires the use of a graded approach and provides guidance for the selection of a degree of complexity for the hazard analysis. Additional requirements and guidance have been provided in this standard (ASCE 43) for performing site-specific probabilistic seismic hazard analysis, site response analysis, and for establishing uniform hazard response spectra (UHRS).

Standards used in the seismic design of nuclear safety-related structures include the following:

- ASCE 1, *Guidelines for Design and Analysis of Nuclear Safety-Related Earth Structures* (ASCE 1982) provides criteria and guidelines to be used in construction of earth structures forming part of the ultimate heat sink or acting to protect nuclear power plant sites from flood, storm surge, or other types of natural or artificial external load phenomena.
- ASCE 4, *Seismic Analysis of Safety-Related Nuclear Structures* (ASCE 2016b) provides minimum requirements and acceptable methods for the seismic analysis of a nuclear facility. ASCE 4 also provides a methodology for calculating seismic response in structures and to derive input motions for use in the seismic design qualification of electrical and mechanical systems and components located in or supported by buildings or other civil structures.
- ASCE 7, *Minimum Design Loads and Associated Criteria for Buildings and Other Structures* (ASCE 2016a) provides minimum design load requirements and procedures to determine the effects of loads for the design of buildings and other civil structures, as well as mechanical and electrical distribution systems and components.

Other codes and standards used in the seismic design of nuclear safety-related structures include the following:

- ACI 349, *Code Requirements for Nuclear Safety-Related Concrete Structures and Commentary* (ACI 2013b) addresses the design and construction of concrete structures that form part of a nuclear safety-related facility. These facilities include concrete structures inside and outside a nuclear reactor containment system, but not concrete reactor vessels or their concrete containment structures (which are defined in a Joint ACI-ASME Committee 359 Code, ACI 2015). ACI 349 is based on ACI 318 (ACI 2014) except that (a) it includes loads and load combinations applicable to nuclear structures, and (b) some of the prescriptive details of ACI 318 for special systems are relaxed in recognition of the smaller nonlinear demands placed on nuclear structures.
- ASME, *Boiler and Pressure Vessel Code*, Section III, "Rules for Construction of Nuclear Facility Components" (ASME 2019a).
- ANSI/AISC N690 2018, *Specification for Safety-Related Steel Structures in Nuclear Facilities* (AISC 2018) applies to the construction of safety-related nuclear steel structures. This specification includes design loads, load combinations, and requirements for treatment of impact and impulse loads; it is otherwise compatible with ANSI/AISC 360 2016, *Specification for Structural Steel Buildings* (AISC 2016c). Only those sections that differ from the ANSI/AISC 360 specification provisions are included in the N690 specification.

Codes and standards used in the design and seismic qualification of mechanical and electrical equipment and distribution systems and components are identified in Chapter 8 of this standard and its associated commentary.

C1.5 ALTERNATIVE METHODS TO MEET THE INTENT OF THIS STANDARD

The DBE ground motion is defined in Equation (2-1) in terms of a design response spectrum (DRS). Structures whose elements are designed for DRS demands using design capacities derived in accordance with referenced materials standards, as modified in this standard, are expected to achieve the target performance goal listed in Table 1-1. The DRS is formulated to achieve both of the following:

1. Less than about a 1% probability of unacceptable performance for the DBE ground motion, and
2. Less than about a 10% probability of unacceptable performance for a ground motion equal to 150% of the DBE ground motion.

Alternative methods that achieve the target performance goal are acceptable.

Seismic fragility functions (i.e., conditional probabilities of unacceptable performance versus seismic demand) are typically assumed to be lognormally distributed so that they can be fully described by two terms: a median and an estimate of the composite variability, β [Equation (C1-11)], a logarithmic standard deviation. The median in this case is a seismic margin factor, $F_{PF=50\%}$, where F_{PF} is the ratio of the acceleration corresponding to a conditional failure probability PF to the DBE acceleration. Satisfying the two target levels of conservatism defined here results in the following seismic margin factors $F_{1\%}$, $F_{5\%}$, $F_{10\%}$, and $F_{50\%}$, corresponding to a 1%, 5%, 10%, and 50% conditional probability of unacceptable performance (i.e., failure), respectively:

β	$F_{1\%}$	$F_{5\%}$	$F_{10\%}$	$F_{50\%}$
0.30	1.10	1.35	1.50	2.20
0.40	1.00	1.31	1.52	2.54
0.50	1.00	1.41	1.69	3.20
0.60	1.00	1.50	1.87	4.04

For a logarithmic standard deviation less than 0.39, the second conditional failure probability controls. For β greater than 0.39, the first goal controls. By satisfying both goals, the following margins are achieved:

$$F_{1\%} \geq 1.0$$
$$F_{5\%} \geq 1.3$$
$$F_{10\%} \geq 1.5$$
$$F_{50\%} \geq 2.2$$

Factors of Safety Achieved by Seismic Acceptance Criteria

Introduction. In this standard, component design strengths are assumed, namely, component nominal strengths according to ACI and AISC multiplied by action-specific strength reduction factors (ϕ), and strength according to ASME Service Level D. Seismic demand is calculated according to ASCE 4, using the seismic hazard of Chapter 2 of this standard. Limit state–

dependent inelastic energy absorption factors can be implemented according to Chapters 5 and 8 to reduce seismic demands for actions associated with ductile response.

Estimation of Median Conservatism Introduced by Standard Seismic Acceptance Criteria. The median seismic capacity, $C_{50\%}$, can be estimated from

$$C_{50\%} = \frac{S_{50\%}}{D_{50\%}} F_{\mu 50\%} \text{DBE} \quad \text{(C1-1)}$$

where

$S_{50\%}$ = Median estimate of the component's seismic strength,
$D_{50\%}$ = Seismic demand for DBE input, and
$F_{\mu 50\%}$ = Inelastic energy absorption factor.

The standard seismic capacity, C_{STD}, is given by

$$C_{\text{STD}} = \frac{S_{\text{STD}}}{D_{\text{STD}}} F_{\mu \text{STD}} \text{ DBE} \quad \text{(C1-2)}$$

where

S_{STD} = Deterministic strength,
D_{STD} = Demand, and
$F_{\mu \text{STD}}$ = Nonlinear factors,

as defined in accordance with this standard. Defining R_S, R_D, and R_N as the median conservatism ratios associated with this standard,

$$S_{50\%} = R_S S_{\text{STD}}$$

$$D_{50\%} = D_{\text{STD}}/R_D$$

$$F_{\mu 50\%} = R_N F_{\mu \text{STD}} \quad \text{(C1-3)}$$

and

$$C_{50\%} = R_C C_{\text{STD}} \quad \text{(C1-4)}$$

$$R_C = R_S R_D R_N \quad \text{(C1-5)}$$

where R_C is the overall median conservatism ratio associated with the acceptance criteria of this standard. The values of R_S, R_D, and R_N are estimated in the following three subsections.

Median Strength Conservatism Ratio. Based on a review of median capacities from past seismic probabilistic risk assessment studies with respect to US code-specified ultimate strengths for a number of failure modes, code *design* strengths have at least a 98% probability of exceedance for ductile failure modes if conservatisms for material strengths, empirical equations, and strain-rate effects are considered. For low-ductility failure modes such as out-of-plane shear in concrete walls and slabs, an additional 1.33 factor is applied to the 98th percentile capacity. This factor accounts for the additional margin provided by standards for the design strength of low-ductility failure modes. Accordingly,

$$\text{(Ductile)} \qquad R_S = e^{2.054\beta_S} \quad \text{(C1-6)}$$

$$\text{(Low ductility)} \qquad R_S = 1.33e^{2.054\beta_S}$$

where β_S is the *strength* logarithmic standard deviation, typically in the range of 0.2 to 0.4, and 2.054 is the standard normal variate for a 2% nonexceedance probability (NEP).

Median Demand Conservatism Ratio. The goal of ASCE 4 is to calculate seismic demands with an 80% probability of nonexceedance conditioned on a design response spectrum. Analysis of multiple soil columns and enveloping the resultant spectra according to Chapter 5 of ASCE 4 is assumed to achieve the 80th percentile. The median demand ratio, R_D, can therefore be estimated as

$$R_D = e^{0.842\beta_D} \quad \text{(C1-7)}$$

where β_D is the *seismic demand* logarithmic standard deviation for a specified seismic input, typically between 0.2 and 0.4 and 0.842 is the standard normal variate for a 20% NEP.

Median Nonlinear Conservatism Ratio. In this standard, the nonlinear factor is aimed at the 5% NEP level. For ductile failure modes such as flexure, the median nonlinear factor ratio, R_N, is

$$\text{(Ductile)} \quad R_N = e^{1.645\beta_N} \quad \text{(C1-8)}$$

where β_N is the logarithmic standard deviation for the *nonlinear* factor, typically between 0.2 and 0.4, and 1.645 is the standard normal variate for a 5% NEP.

For low-ductility (brittle) failure modes, no credit is taken for a nonlinear factor, namely,

$$\text{(Low ductility)} \qquad R_N = 1.0 \quad \text{(C1-9)}$$

Capacity Conservatism. Combining Equations (C1-5) through (C1-8), the median capacity ratio, R_C, is

$$\text{(Ductile failures)} \qquad R_C = e^{2.054\beta_S + 0.842\beta_D + 1.645\beta_N} \quad \text{(C1-10)}$$

$$\text{(Low ductility)} \qquad R_C = 1.33e^{2.054\beta_S + 0.842\beta_D}$$

and

$$C_{1\%} = R_C C_{\text{STD}} e^{-2.326\beta} \quad \text{(C1-11)}$$

where β is a composite variability given by

$$\beta = [\beta_S^2 + \beta_D^2 + \beta_N^2]^{1/2} \quad \text{(C1-12)}$$

The resulting nominal factor of safety, $F_{N1\%}$, against a 1% conditional probability of failure is

$$F_{N1\%} = \frac{C_{1\%}}{C_{\text{STD}}} = R_C e^{-2.326\beta} \quad \text{(C1-13)}$$

and the nominal factor of safety, $F_{N10\%}$, against a 10% conditional probability of failure is

$$F_{N10\%} = \frac{C_{10\%}}{C_{\text{STD}}} = R_C e^{-1.282\beta} \quad \text{(C1-14)}$$

Table C1-1 presents $F_{N1\%}$ for typical values of β_S, β_D, and β_N. Over this range of values,

$$F_{N1\%} \approx 1.0 \quad \text{(C1-15)}$$

Table C1-1. Nominal Factor of Safety, $F_{N1\%}$.

Strength Variability (β_S)	Demand Variability (β_D)	Low-Ductility Failure Modes	Ductile Failure Modes $\beta_N = 0.2$	Ductile Failure Modes $\beta_N = 0.4$
0.2	0.2	1.23	1.11	1.10
	0.3	1.12	1.03	1.07
	0.4	1.00	0.94	1.01
0.3	0.2	1.26	1.17	1.21
	0.3	1.18	1.11	1.19
	0.4	1.08	1.03	1.13
0.4	0.2	1.27	1.20	1.29
	0.3	1.22	1.16	1.27
	0.4	1.14	1.10	1.23

with $F_{N1\%}$ ranging between 0.94 and 1.29, with a median value of 1.14. Table C1-2 presents $F_{N10\%}$ for typical values of β_S, β_D, and β_N. Similarly,

$$F_{N10\%} > 1.5 \tag{C1-16}$$

Accordingly, both performance statements 1 and 2 given in Section C1.5 are met.

Table C1-2. Nominal Factor of Safety, $F_{N10\%}$.

Strength Variability (β_S)	Demand Variability (β_D)	Low-Ductility Failure Modes	Ductile Failure Modes $\beta_N = 0.2$	Ductile Failure Modes $\beta_N = 0.4$
0.2	0.2	1.66	1.59	1.84
	0.3	1.63	1.59	1.88
	0.4	1.57	1.57	1.89
0.3	0.2	1.80	1.80	2.12
	0.3	1.82	1.82	2.18
	0.4	1.81	1.81	2.20
0.4	0.2	2.00	2.00	2.41
	0.3	2.04	2.04	2.49
	0.4	2.05	2.05	2.53

CHAPTER C2
EARTHQUAKE GROUND MOTION

C2.1 SEISMIC HAZARD EVALUATION

The products of a probabilistic seismic hazard assessment (PSHA) are seismic hazard curves and uniform hazard response spectra (UHRS) associated with several hazard exceedance frequencies. Example seismic hazard curves are shown in Figure C2-1(a). Earthquake input excitation to be used for design by these provisions is defined by a UHRS shape, such as that shown in Figure C2-1(b).

C2.2 DEVELOPMENT OF DESIGN BASIS EARTHQUAKE GROUND MOTIONS

The 2005 edition of ASCE 43 defined the design response spectrum (DRS) using a UHRS and a design factor (DF) greater than 1.0. This edition of ASCE 43 defines the DRS using a UHRS and a scale factor (SF) less than 1.0. The net effect of this change is small, as described in this section. The return periods used to determine the UHRS in ASCE 43-05 and in this standard are different, also as described in this section.

Technical Basis for the Scale Factor. The equations presented in this standard for defining the DRS differ in two ways from the equations presented in ASCE 43-05. The 2005 edition of ASCE 43 and Kennedy (2011) provide the technical basis for the equations of ASCE 43-05.

ASCE 43-05 introduced a probability ratio, R_P, which was defined as the hazard exceedance probability, H_D, divided by the target performance goal, P_F. The ratio was set equal to 4, 10, and 10 for Seismic Design Categories (SDCs) 3, 4, and 5, respectively. The target performance goals in ASCE 43-05 for SDC 3, 4, and 5 were 1×10^{-4}, 4×10^{-5}, and 1×10^{-5}, respectively. These target performance goals are retained in this revision of the standard. Seismic Design Category 2 was added to ASCE 43-19, with a target performance goal of 4×10^{-4}.

Early deliberations regarding the introduction of SDC 2 into ASCE 43-19 addressed the appropriate value for R_P, which was anticipated to be less than 4 (the value used for SDC 3). Calculations by Kennedy (2011) indicated that the equation to define DF for $R_P = 2$ would have to differ from that used for $R_P = 4$ and 10 (Equations (2-3a) and (2-3b) in ASCE 43-05). Furthermore, during the investigation to define DF for $R_P = 2$, a problem with the use of $R_P = 4$ for SDC 3 was identified. First, it was observed that the portion of the hazard curve that dominated the computed probability of failure (or target performance goal), P_F, ranged from about $H_D = 10 \times P_F$ to $H_P = P_F$. The use of $R_P = 10$ preserved this range when defining the slope factor, A_R. The use of $R_P = 4$ shifted the range over which A_R was defined from $H_D = 4 \times P_F$ to $0.1H_D = 0.4 \times P_F$, which was not the dominant range contributing to P_F. Shifting the range on the hazard curve over which A_R is defined typically results in a small increase in A_R. Although the small increase in A_R is compensated by the way in which the DF is computed, it was judged to be fundamentally more correct to use $R_P = 10$ for SDC 3. For typical seismic hazard curves, this revision resulted in less than a 5% change in the computed DRS for SDC 3.

In this revision of ASCE 43, the DRS is computed by multiplying the UHRS_{H_P}, defined at the exceedance frequency of the target performance goal (i.e., $H_P = P_F$) by SF, where SF is always less than unity. In ASCE 43-05, for $R_P = 10$, the DRS was computed by multiplying the UHRS_{H_D}, defined at the exceedance frequency $H_D = 10 \times P_F$, by DF, where DF was always greater than unity. This change is made for the following three reasons:

1. The DF approach of ASCE 43-05 masked the relationship between the DRS and UHRS_{H_P}. Because of the conservatism associated with the design criteria in ASCE 43, the SF rightfully must be less than unity and must reduce with increasing A_R. This trend was not clearly obvious with the DF approach of ASCE 43-05.
2. The SF is much less sensitive to A_R than the former DF over the most likely range of A_R, namely, 2.0 to 4.0. The SF reduces by only 15%, whereas DF increases by a factor of 1.75 as A_R changes from 2.0 to 4.0.
3. For reasons discussed subsequently, a practical lower bound must be placed on SF for large values of A_R. This practical bound is more easily imposed on SF than on DF.

With the exception of a few sites of nuclear facilities with high A_R, for SDC 4 and SDC 5, the ordinates of the DRS computed from Equations (2-1) through (2-3) of this standard are identical to those computed using the corresponding equations in ASCE 43-05, which is demonstrated here.

Derivation of SF for $R_P = 10$. ASCE (2005) and Kennedy (2011) each summarize the derivation of the DF for $R_P = 10$, as implemented in ASCE 43-05:

$$\text{DRS} = \text{DF} \times \text{UHRS}_{H_D} \quad \text{(C2-1)}$$

where $H_D = 10 \times P_F$ is the mean annual exceedance frequency at which the UHRS_{H_D} is defined, and

$$\text{DF} = \text{Maximum}\ [\text{DF}_1;\ \text{DF}_2] \quad \text{(C2-2a)}$$

$$\text{DF}_1 = 1.0 \quad \text{(C2-2b)}$$

$$\text{DF}_2 = 0.6{A_R}^{0.8} \quad \text{(C2-2c)}$$

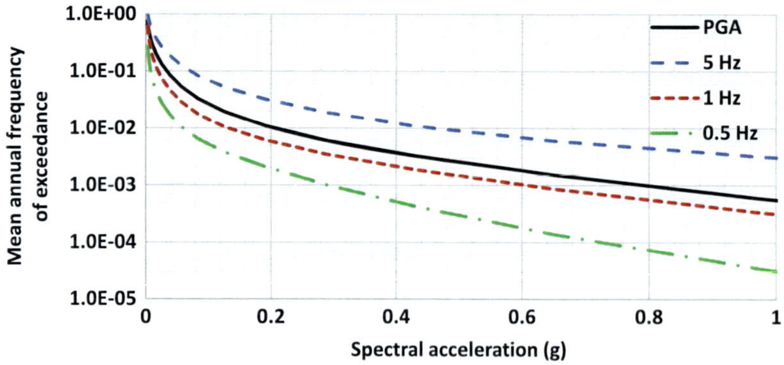

(a) Seismic hazard curves

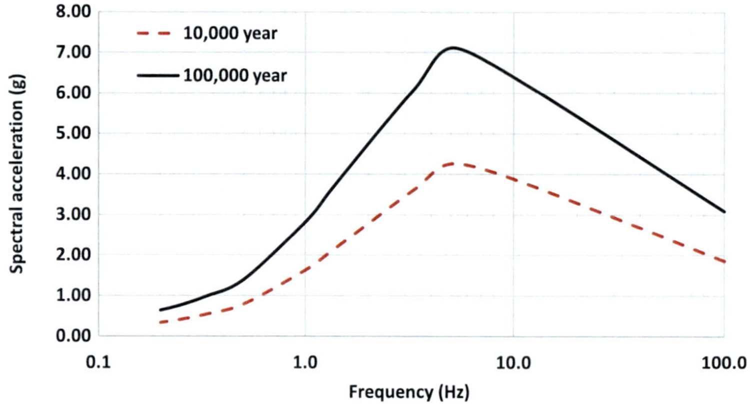

(b) Uniform hazard response spectra

Figure C2-1. Definitions of earthquake input excitation.

In Section 2.2 of this standard, the DRS is defined as

$$DRS = SF \times UHRS_{H_P} \quad (C2\text{-}3)$$

where $H_P = P_F$ is the mean annual exceedance frequency at which the UHRS is defined. In both editions of the standard,

$$A_R = \frac{UHRS_{H_P}}{UHRS_{H_D}} \quad (C2\text{-}4)$$

Combining Equations (C2-1), (C2-3), and (C2-4),

$$SF = \frac{DF}{A_R} \quad (C2\text{-}5)$$

and from Equation (C2-2a),

$$SF = \text{Maximum } [SF_1;\ SF_2] \quad (C2\text{-}6a)$$

$$SF_1 = A_R^{-1} \quad (C2\text{-}6b)$$

$$SF_2 = 0.6A_R^{-0.2} \quad (C2\text{-}6c)$$

The intensity of the ground motion at some sites of low seismic hazard is very small at the $H_D = 10 \times P_F$ annual frequency of exceedance, with A_R greater than 4.5. These hazard curves cannot generally be approximated as linear in log space between the H_D and H_P exceedance frequencies, and the calculated value of A_R is unreliable. To avoid such problems, a minimum value of SF is given by

$$SF_3 = 0.45 \quad (C2\text{-}6d)$$

The first term, SF_1, controls when $A_R < 1.89$. The third term, SF_3, controls when $A_R > 4.21$. The second term, SF_2, controls between these two limiting values of A_R. The terms SF_1, SF_2, and SF_3 comprise an empirical curve that closely matches the exact SF solution (assuming linear hazard) for $\beta = 0.35$ and A_R between 2 and 4.

Table C2-1 lists SF as a function of A_R.

Alternate Derivation of SF for $R_P = 10$. The purpose of the previous subsection was to show the equivalency of SF and DF. Below, the original form of the derivation of DF, as presented in ASCE (2005) is repeated herein for SF and $R_P = 10$.

Figure C2-2 presents normalized seismic hazard curves for two regions expressed in terms of mean annual frequency of exceedance versus normalized spectral acceleration. Spectral accelerations are normalized to their value at an exceedance frequency of 1×10^{-4} to better compare the hazard curve slopes. The curve labeled "Eastern U.S." is representative of Central and Eastern United States (CEUS) sites. The California site is near Santa Maria, California, which has high-frequency ground motion dominated by active nearby faults and low-frequency ground motion dominated by large events on the more distant San Andreas Fault.

Several general observations can be made for the normalized hazard curves shown in Figure C2-2:

Table C2-1. Scale Factor as a Function of A_R.

A_R	SF
1.50	0.667
1.75	0.571
2.00	0.522
2.25	0.510
2.50	0.500
2.75	0.490
3.00	0.482
3.25	0.474
3.50	0.467
3.75	0.461
4.00	0.455
4.25	0.450
4.50	0.450

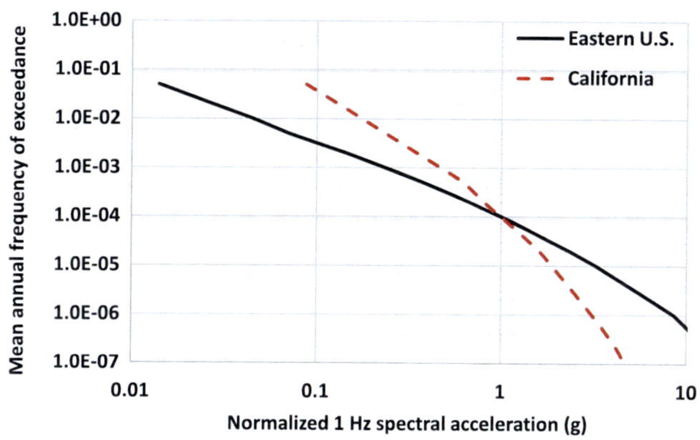

(a) 1 Hz

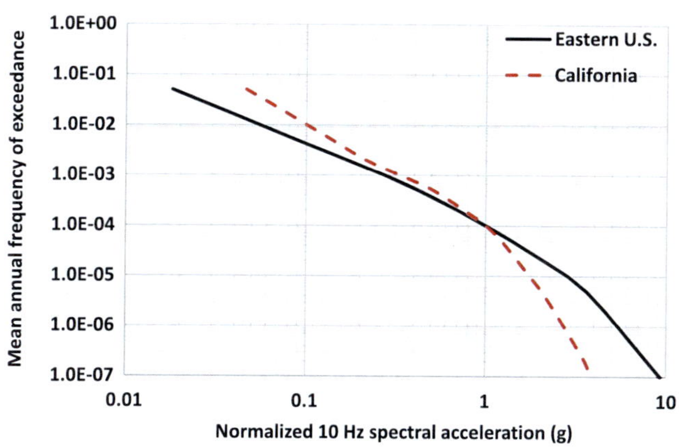

(b) 10 Hz

Figure C2-2. Normalized seismic hazard curves.

1. The hazard curve slopes are much steeper for the California site than for the Eastern U.S. site.
2. When plotted on a log-log plot, the hazard curves can be approximated as being linear over any 10-fold change in exceedance frequency.

Based on these observations, ASCE (2005) and Kennedy (2011) enabled the writing of risk-consistent seismic design criteria for sites with widely varying hazard curve slopes. This approach has been modified here to define a DRS consistent with uniform risk in terms of a mean hazard response spectrum $UHRS_{H_P}$ defined as the performance goal probability, P_F, and SF:

$$\mathrm{DRS} = \mathrm{SF} \times \mathrm{UHRS}_{H_P} \tag{C2-7}$$

Typical seismic hazard curves are close to linear when plotted on a log-log scale and so can be approximated by a power law over any 10-fold (at least) change in exceedance frequency:

$$H(a) = K_1 a^{-K_H} \tag{C2-8}$$

where $H(a)$ is the annual frequency of exceedance of ground motion level a, K_1 is a constant, and K_H is a slope parameter defined by

$$K_H = \frac{1}{\log(A_R)} \tag{C2-9}$$

where A_R is the ratio of spectral acceleration response corresponding to a 10-fold reduction in exceedance frequency.

There is a closed-form solution for the seismic risk if the fragility curve describing damage is lognormally distributed, and the hazard curve is defined by Equation (C2-8). This closed-form solution is derived in Appendix A of Kennedy (2011) and elsewhere to be

$$P_F = H(F_{50\%}{}^{-K_H}) \times e^{a} \tag{C2-10}$$

where

$$F_{50\%} = \frac{C_{50\%}}{C_H} \tag{C2-11}$$

and

$$a = \frac{1}{2}(K_H\beta)^2 \tag{C2-12}$$

where H is any reference exceedance frequency, C_H is the UHRS ground motion level that corresponds to exceedance frequency H from the seismic hazard curve, $C_{50\%}$ is the median seismic capacity, and β is the natural logarithmic standard deviation of the fragility curve.

Equation (C2-10) is often referred to as the *seismic risk equation*. The only approximations in its derivation are (1) the hazard curve is approximated by Equation (C2-8) over the exceedance frequency range of interest, and (2) the fragility curve is lognormally distributed.

When H is defined by $H_P = P_F$, from Equation (C2-10) the required factor of safety $F_{50\%}$ to achieve the performance goal P_F becomes

$$F_{50\%} = e^{K_H\beta^2} \tag{C2-13}$$

Table C2-2 lists $F_{50\%}$ as a function of A_R and β.

The seismic acceptance criteria specified in this standard assume that both of the following nominal factors of safety, F_N, are achieved:

$F_{N1\%} \geq 1.0$ against a 1% conditional probability of failure (C2-14a)

Table C2-2. Required Safety Factor $F_{50\%}$ as a Function of A_R and β.

	$F_{50\%}$			
A_R	β = 0.3	β = 0.4	β = 0.5	β = 0.6
1.50	1.29	1.58	2.03	2.78
1.75	1.20	1.39	1.67	2.10
2.00	1.16	1.30	1.51	1.82
2.25	1.14	1.26	1.43	1.67
2.50	1.12	1.22	1.37	1.57
2.75	1.11	1.20	1.33	1.51
3.00	1.10	1.18	1.30	1.46
3.25	1.09	1.17	1.28	1.42
3.50	1.09	1.16	1.26	1.39
3.75	1.08	1.15	1.24	1.37
4.00	1.08	1.14	1.23	1.35
4.25	1.07	1.14	1.22	1.33
4.50	1.07	1.13	1.21	1.32
4.75	1.07	1.13	1.20	1.30
5.00	1.07	1.12	1.20	1.29
5.25	1.06	1.12	1.19	1.28
5.50	1.06	1.11	1.18	1.28
5.75	1.06	1.11	1.18	1.27
6.00	1.06	1.11	1.17	1.26

Table C2-3. Nominal Factors of Safety.

β	$F_{N1\%}$	$F_{N10\%}$	$F_{N50\%}$
0.3	1.1	1.50	2.20
0.4	1.0	1.52	2.54
0.5	1.0	1.69	3.20
0.6	1.0	1.87	4.04

$F_{N10\%} \geq 1.5$ against a 10% conditional probability of failure (C2-14b)

Section C1.3 provides the justification that both of these nominal factors of safety are achieved. The requirements of Equations (C2-14a) and (C2-14b) achieve the nominal median factors of safety, $F_{N50\%}$, of Table C2-3.

The scale factor is then given by

$$\mathrm{SF} = F_{50\%}/F_{N50\%} \quad \text{(C2-15)}$$

where $F_{50\%}$ is given by Equation (C2-13) and $F_{N50\%}$ is given in Table C2-3. Values for the scale factor are listed in Table C2-4 as a function of A_R and β. The values of SF calculated using Equation (C2-6a) are also shown in the table. In general, the values of SF from Equation (C2-6a) are conservatively biased.

Demonstration that the Scale Factor Approach of ASCE 43-19 Reasonably Achieves the Specified Target Performance Goals, P_F. To demonstrate that the SF approach to determining the DRS reasonably achieves the specified target performance goal, the Eastern US and California hazard curves, shown in Figure C2-2 for 10 Hz and 1 Hz, are used for calculations. Normalized spectral

Table C2-4. Required Scale Factor as a Function of A_R and β.

	SF				
A_R	β = 0.3	β = 0.4	β = 0.5	β = 0.6	Eq. (C2-6a)
1.50	0.586	0.621	0.636	0.688	0.667
1.75	0.546	0.548	0.523	0.520	0.571
2.00	0.527	0.514	0.473	0.450	0.522
2.25	0.516	0.495	0.446	0.413	0.510
2.50	0.508	0.482	0.428	0.389	0.500
2.75	0.503	0.472	0.415	0.373	0.490
3.00	0.499	0.466	0.406	0.361	0.482
3.25	0.496	0.461	0.399	0.352	0.474
3.50	0.493	0.457	0.393	0.345	0.467
3.75	0.491	0.453	0.389	0.339	0.461
4.00	0.489	0.450	0.385	0.334	0.455
4.25	0.488	0.448	0.381	0.330	0.450
4.50	0.486	0.446	0.378	0.326	0.450
4.75	0.485	0.444	0.376	0.323	0.450
5.00	0.484	0.442	0.374	0.320	0.450
5.25	0.483	0.441	0.372	0.318	0.450
5.50	0.482	0.439	0.370	0.316	0.450
5.75	0.482	0.438	0.368	0.314	0.450
6.00	0.481	0.437	0.367	0.312	0.450

Table C2-5. Normalized Spectral Accelerations.

	Eastern United States		California	
$H_{(SA)}$	1 Hz	10 Hz	1 Hz	10 Hz
5×10^{-2}	0.014	0.018	0.087	0.046
2×10^{-2}	0.027	0.034	0.13	0.072
1×10^{-2}	0.045	0.055	0.175	0.100
5×10^{-3}	0.07	0.089	0.236	0.139
2×10^{-3}	0.143	0.169	0.351	0.215
1×10^{-3}	0.235	0.275	0.474	0.334
5×10^{-4}	0.383	0.424	0.629	0.511
2×10^{-4}	0.681	0.709	0.814	0.762
1×10^{-4}	1.00	1.00	1.00	1.00
5×10^{-5}	1.46	1.41	1.23	1.22
2×10^{-5}	2.35	2.13	1.61	1.51
1×10^{-5}	3.27	2.88	1.89	1.76
5×10^{-6}	4.38	3.65	2.20	2.05
2×10^{-6}	6.44	4.62	2.68	2.42
1×10^{-6}	8.59	5.43	3.10	2.72
5×10^{-7}	10.34	6.38	3.58	3.06
2×10^{-7}	13.21	7.90	4.24	3.56
1×10^{-7}	15.90	9.28	4.67	3.84

acceleration values, SA, versus annual exceedance frequencies, $H_{(SA)}$, for the four hazard curves considered, are shown in Table C2-5.

For each of these four hazard curves and for SDC 2, 3, 4, and 5, each of the following is tabulated in Table C2-6:

Table C2-6. Comparison of Achieved Performance Probabilities, P_F, versus Target Goals for Various Cases.

SDC	Target Prob. Performance Goals (×10^{-5})	Hazard Curve	$UHRS_{H_P}$	A_R	SF	DRS	Achieved Performance Probabilities P_F (×10^{-5}) β = 0.3	β = 0.4	β = 0.5	β = 0.6
2	40.0	EUS 1 Hz	0.441	5.29	0.45	0.198	44.60	38.7	29.7	23.4
		EUS 10 Hz	0.481	4.62	0.45	0.216	45.70	39.3	29.7	23.1
		Calif. 1 Hz	0.670	2.58	0.496	0.333	41.10	35.2	25.2	19.5
		Calif. 10 Hz	0.563	3.63	0.464	0.261	43.50	36.6	26.6	20.1
3	10.0	EUS 1 Hz	1.00	4.26	0.45	0.450	11.60	10.0	7.60	5.90
		EUS 10 Hz	1.00	3.64	0.463	0.463	11.40	9.80	7.30	5.60
		Calif. 1 Hz	1.00	2.11	0.517	0.517	10.70	9.50	6.90	5.40
		Calif. 10 Hz	1.00	2.99	0.482	0.482	10.70	8.90	6.20	4.70
4	4.0	EUS 1 Hz	1.64	3.72	0.461	0.76	4.50	3.90	2.90	2.30
		EUS 10 Hz	1.56	3.25	0.474	0.74	4.40	3.80	2.80	2.10
		Calif. 1 Hz	1.31	1.86	0.524	0.69	4.00	3.70	2.70	2.20
		Calif. 10 Hz	1.28	2.28	0.509	0.65	4.30	3.60	2.60	2.00
5	1.0	EUS 1 Hz	3.27	3.27	0.473	1.55	1.09	0.93	0.69	0.52
		EUS 10 Hz	2.88	2.88	0.486	1.40	1.03	0.87	0.62	0.46
		Calif. 1 Hz	1.89	1.89	0.529	1.00	1.04	0.96	0.73	0.61
		Calif. 10 Hz	1.76	1.76	0.568	1.00	0.84	0.78	0.58	0.48

1. Target probabilistic performance goal;
2. $UHRS_{H_p}$ spectral acceleration, SA_{UHRS}, defined at the H_P hazard exceedance frequency;
3. Slope ratio, A_R, computed from Equation (C2-4);
4. Scale factor, SF, computed from Equation (C2-6a);
5. Resulting DRS spectral acceleration from Equation (C2-1); and
6. Resulting actually achieved unacceptable performance probabilities, P_F, obtained by numerical integration of hazard curves and fragility curves for β = 0.3, 0.4, 0.5, and 0.6.

The ratios of the achieved to target performance probabilities, as shown in Table C2-7, range from 1.16 to 0.46, with a mean ratio of 0.80. The specified SF values coupled with the dual acceptance criteria specified achieve a P_F no more than 1.16 (unconservative) times the target. The penalty is that the achieved P_F might be as low as 0.46 (conservative) times the target. This variability in the achieved P_F cannot be avoided for any simple criteria because P_F varies by about a factor of two as a function of β. The goal has been to specify SF values that accurately achieve the target performance goal for low-variability failure modes (β between 0.3 and 0.4) while accepting increased conservatism for larger variability failure modes (β larger than 0.4).

Table C2-7. Ratio of Achieved to Target Performance Probabilities.

SDC	Hazard Curve	Ratio of Achieved to Target P_F β = 0.3	β = 0.4	β = 0.5	β = 0.6
2	EUS 1 Hz	1.12	0.97	0.74	0.58
	EUS 10 Hz	1.14	0.98	0.74	0.58
	Calif. 1 Hz	1.03	0.88	0.63	0.49
	Calif. 10 Hz	1.09	0.92	0.66	0.50
3	EUS 1 Hz	1.16	1.00	0.76	0.59
	EUS 10 Hz	1.14	0.98	0.73	0.56
	Calif. 1 Hz	1.07	0.95	0.69	0.54
	Calif. 10 Hz	1.07	0.89	0.62	0.47
4	EUS 1 Hz	1.12	0.98	0.72	0.58
	EUS 10 Hz	1.10	0.95	0.70	0.52
	Calif. 1 Hz	1.00	0.92	0.68	0.55
	Calif. 10 Hz	1.08	0.90	0.65	0.50
5	EUS 1 Hz	1.09	0.93	0.69	0.52
	EUS 10 Hz	1.03	0.87	0.62	0.46
	Calif. 1 Hz	1.04	0.96	0.73	0.61
	Calif. 10 Hz	0.84	0.78	0.58	0.48

C2.3 METHOD TO DEFINE DESIGN RESPONSE SPECTRA AT VARIOUS DEPTHS IN THE SITE PROFILE

The calculation procedure starts with mean hazard curves defined by PSHA as outcrop motions located at bedrock or some other reference horizon. Site amplification functions (SAF) are developed by convolving the deaggregated rock response spectra obtained from the rock hazard curves, defined as annual frequencies of exceedance (AFEs) of rock acceleration as a function of spectral frequency (SA^R, spectral acceleration, rock). Repeating the site response analysis for different levels of rock hazard AFE, the mean hazard curves in the site profile (SA^S, spectral acceleration, soil) at the depth(s) of interest can be obtained. These hazard curves can then be used to develop uniform hazard response spectra (UHRS) in the site profile at the depth of interest. SFs at each spectral frequency are computed as described in Section 2.2 using the hazard curves at two different hazard levels with a 10-fold reduction in exceedance frequency and multiplied by the corresponding UHRS to calculate DRS at the locations of interest.

The steps to develop hazard curves within the site profile are implemented using a number of approaches described in

McGuire et al. (2001), designated as Approaches 1, 2A, 2B, and 3; higher numbers are associated with the more rigorous approaches specifically with respect to the potential sensitivity of the site amplification functions to magnitude and distance dependency of seismic sources and nonlinearity of the soil properties and consideration of uncertainty in the site profile and dynamic soil properties.

In the following paragraphs, the more rigorous of the approximate approaches, Approach 3, is described first. Then Approaches 2B, 2A, and 1, which include increasingly simplified assumptions compared to Approach 3, are described in decreasing order of computational rigor. The analyst should justify the use of Approaches 1, 2A, and 2B by considering the potential nonconservatism of the approach compared to the results expected from the more complete Approach 3.

Approach 3 incorporates the site amplification functions into the hazard calculation through convolution of the bedrock hazard curves (SA^R) for each spectral frequency (period) with the probability density function for the SAFs to compute hazard curves for locations within the site profile. Convolution permits each bedrock ground motion level to contribute to the hazard for each ground motion level (SA^S) at the location of interest in the site profile. Thus, the hazard within the site profile, $G(z)$ is calculated as

$$G(z) = \sum_{x_j} P\left[\text{SAF} > \frac{z}{x_j}\middle| x_j, m_i, d_i\right] \times P[\text{SA}^R = x_j] \qquad \text{(C2-16)}$$

in which z is the ground acceleration level (SA^S) in the site profile, $P[\text{SAF} > \frac{z}{x_j}|x_j, m_i, d_i]$ is the probability that SAF is greater than z/x_j given that $\text{SA}^R = x_j$, m_i, and d_i are source, magnitude, and distance associated with the hazard at rock at the spectral frequency of interest. The annual probability of occurrence for $\text{SA}^R = x_j$ is $P[\text{SA}^R = x_j]$, where $\text{SA}^R = x_j$ can be approximated by differencing the rates from the computed bedrock hazard curve about $\text{SA}^R = x_j$. Through this equation, all levels of bedrock motion contribute to the hazard at the location of interest within the site profile.

Assuming SAF is lognormally distributed, $P[\text{SAF} > \frac{z}{x_j}|x_j, m_i, d_i]$ is computed as

$$\sum_{x_j} P\left[\text{SAF} > \frac{z}{x_j}\middle| x_j, m_i, d_i\right] = \hat{\Phi}\left(\frac{\ln\left[\frac{z}{x}\right] - \mu_{\ln \text{AF}|x,m,d}}{\sigma_{\ln \text{AF}|x,m,d}}\right) \qquad \text{(C2-17)}$$

where $\mu_{\ln \text{AF}|x,m,d}$ is the mean and $\sigma_{\ln \text{AF}|x,m,d}$ is the standard deviation of the natural logarithm of SAF, given $\text{SA}^R = x$ conditioned on the magnitude, M, and distance, D, of the characteristic earthquakes associated with x_j, and $\hat{\Phi}$ is the standard complementary Gaussian cumulative distribution function. The mean and standard deviation, $\mu_{\ln \text{AF}|x,m,d}$ and $\sigma_{\ln \text{AF}|x,m,d}$, respectively, are obtained from site response analyses, which are typically performed using sets of 60 or more site response computations, in which a Monte Carlo approach is used to randomly select properties of individual soil layers for which bedrock motions, x_j, that have spectral shapes appropriate for the magnitude and distance associated with the spectral frequencies of interest, are used as input to the site profile. The seismic hazard should be deaggregated to determine mean magnitudes and distances associated with the spectral frequencies considered and the bedrock hazard curve accelerations.

In performing such calculations, information, including means and their uncertainty, is required on the following parameters and their uncertainties for each site soil layer: low strain shear wave velocity (V_s), strain-dependent shear modulus reduction and hysteretic damping properties, soil density, and layer thickness. Corresponding information is required on total depth of the profile to bedrock. This set of parameters assumes that the equivalent linear method of site response is used. The list of parameters and their uncertainty associated with other approaches to site response may be different.

A number of simplifications to Approach 3 are often made in implementation, such as neglecting the relationship of source distance to SAF, using a limited number of magnitudes to develop SAF, and considering a limited range of SA^R in computing SAF when the site is considered to be primarily linear. Also, additional simplification may be achieved by limiting the set of spectral frequencies considered in developing SAF (e.g., SAF developed for PGA bedrock hazard only and applied at all other frequencies). These simplifications, when used, should be justified. In general, it is preferred to interpret the rock hazard curves for additional frequencies and perform site response analysis for the full range of frequencies of interest.

Approaches 2A and 2B are simplifications of Approach 3 in that the convolution of the bedrock hazard curve is not involved. In these two approaches, a bedrock UHRS ($UHRS^R$) is developed from the bedrock hazard curves for the desired annual frequency of exceedance (AFE). The ordinates of the UHRS are developed at the selected level in the site profile ($UHRS^S$) by propagating the $UHRS^R$ up the site profile using bedrock input motions that consider magnitude, distance, and spectral frequency sensitivity of the SAF.

In performing the site response calculations for Approaches 2A and 2B, the seismic hazard is deaggregated to determine mean magnitudes and distances associated with frequencies of 1 Hz and 10 Hz. These magnitude and distance (M and D) values are used to generate spectral shapes that are then scaled to the UHRS at 1 Hz and 10 Hz, respectively. If the envelope spectrum from these two M and D characteristic events does not fall below the UHRS by more than 10% at any one frequency, the site amplification factors can then be generated from sets of convolution calculations generated from these two controlling events. If the envelope spectrum does fall more than 10% below the $UHRS^R$ at any one frequency, a third characteristic event may need to be incorporated, using a spectral shape scaled to the UHRS at the intermediate frequency. This spectral shape should be appropriate for the characteristic event appropriate for the hazard at the intermediate frequency. If the spectra associated with these controlling events do not differ significantly, a single set of convolutions may be performed using a single broadband spectrum enveloping the UHRS and the two (or more) scaled spectra. Examples of scaling appropriate spectral shapes to the UHRS are presented in McGuire et al. (2001 and 2002). In Approach 2A, single magnitude events are considered at each frequency (e.g., 1 Hz and 10 Hz), whereas in Approach 2B, a range of magnitudes are considered at each frequency.

It should be noted that the selection of characteristic events associated with the UHRS at bounding frequencies of 1 Hz and 10 Hz is considered appropriate for relatively stiff structures typical of reactor systems. For softer structures, structures situated on or in soft soil sites, or sites where the peak of input motion is higher than 10 Hz, the appropriate frequencies to consider as bounding frequencies may be modified somewhat to ensure that an appropriate range of potentially damaging events is considered.

Approach 1 is a further simplification of Approaches 2A and 2B in that the magnitude, distance, and spectral frequency dependence of the SAF are not considered. In Approach 1, the bedrock $UHRS^R$ at the desired AFE is used as input to the site profile and $UHRS^S$ values are computed at the desired location in the site profile.

Caution needs to be exercised when performing such sets of convolution calculations using the bedrock UHRS or a broadband envelope spectrum, particularly for deeper or softer soil sites. This caution involves a concern with the possibility of overdriving the soil column when using synthetic ground motions whose spectrum significantly exceeds the target bedrock outcrop spectrum associated with the characteristic events. By using motions that are more energetic than needed to envelop the targets, calculated soil strains may become unrealistically large, leading to potentially significant increases in computed soil damping and modulus reduction. For deeper sites, this overstraining can lead to shifts in soil column frequencies, which can then influence computed site amplification factors.

Calculation of the SF requires that the slope of the hazard curves be computed. This step is accomplished by computing the $UHRS^S$ associated with each of two bedrock outcrop $UHRS^R$ spectra, at two return periods a factor of 10 apart in annual frequency of exceedance (AFE) using one of the approaches described earlier. After generating the mean $UHRS^S$, the SF is obtained as described in Section 2.3 using the equations in Section 2.2. The DRS is determined by applying the SF to the $UHRS^S$ at the AFE of interest. The intent of the evaluation is to arrive at a risk-consistent DRS at the depth of interest that is consistent with the bedrock outcrop spectra and that can be used as input to the seismic analyses of (near) surface founded structures.

C2.4 CRITERIA FOR DEVELOPING SYNTHETIC OR MODIFIED RECORDED ACCELERATION TIME SERIES

The objective of the development of synthetic or modified recorded acceleration time series is to generate response spectra that closely match or envelop a given target spectrum and that, at the same time, possess characteristics that are appropriate for the tectonic environment. In addition, the time series must have adequate power at all frequencies considered important to site response, soil–structure interaction (SSI), or structural response evaluations. As described in Section 2.3, the output from a seismic hazard evaluation typically used in these types of studies consists of a UHRS appropriate for the annual frequency of exceedance together with characteristic events magnitude and distance (*M*/*D* pairs) controlling at given response frequencies. Details on how such information can be determined from the seismic hazard analysis are described in NRC (1997).

Recordings are typically organized into *M*/*D* bins and applicable geological conditions. The first step is to identify the appropriate *M*/*D* bins that affect the site DRS. Depending on the procedure used in determining the DRS, the dominant contributing events can be selected from either deaggregation of the seismic hazard at important frequencies of interest for the structural analysis or from events appropriate at the displacement, velocity, and acceleration segments of the response spectrum. If a probabilistic seismic hazard analysis is used to define the DRS, these characteristic events can be selected as dominating the hazard at low frequencies (typically defined at 1 to 2.5 Hz), at an intermediate frequency (5 to 10 Hz), and at higher frequencies for CEUS rock sites (around 30 Hz). Alternatively, for all identified credible seismic events, the appropriate *M/D* bins can be selected that characterize the displacement, velocity, and acceleration segments of the response spectrum.

Based on empirical data, information is available (McGuire et al. 2001) on typical properties of events that have been categorized into *M/D* bins, including the effects of site characteristics on these properties. These properties include ranges of strong motion duration, peak velocity (*V*), displacement (*D*), and acceleration (*A*), and ratios of V/A and AD/V^2. Because these parameters are considered important in structural and site evaluation problems, the modified recorded acceleration time series should be used for site-specific analyses. If the target response spectrum is dominated by one *M/D* event, a single set of three-component modified recorded acceleration time series can be developed with each component record obtained from a seed representing that *M/D* event. However, if the target response spectrum contains contributions from multiple *M/D* events, several sets of three-component modified recorded acceleration time series should be developed from seeds reasonably representing the dominant *M/D* bins across the frequency range of engineering interest to the analysis. Each developed acceleration time series to represent a given *M/D* pair should have similar properties as the empirical data in the appropriate *M/D* bins.

Once the recorded motions are selected, the individual recorded ground motion needs to be modified such that its response spectrum matches the DRS over the frequency range of interest. In generating modified recorded acceleration time series, one procedure typically used is to select an appropriate Fourier phase spectrum representing the motion and then modify the corresponding Fourier amplitude spectrum iteratively until the response spectrum of the motion closely envelops the target spectrum. The modified acceleration time series can then be used as input to the analysis models. Time domain methods for adjusting the seed record to produce an appropriate modified record are also acceptable.

When the DRSs are defined as broadband spectra, and it is desired to use individual time series that have spectra consistent with the full range of frequencies of interest, then modified recorded acceleration time series should be developed using phase spectra selected from the available database of recorded events at low (1 to 2.5 Hz), mid (5 to 10 Hz) and high (20 to 30 Hz) frequencies of interest. When the characteristic events (*M/D* pairs of the dominant events) at these frequencies of interest are different, the result may lead to multiple sets of time series based on using a seed from a recorded event from each of the *M/D* bins selected as appropriate at the dominant frequency.

The phase spectrum can be randomly selected or selected from the available database of recorded events. It is generally preferred that phase spectra be selected from the available database, which then results in acceleration, velocity, and displacement time series that seem reasonable (i.e., they resemble actual earthquake recordings from the appropriate *M/D* bin). When trying to represent near-field events where unusual characteristics of the recorded motions have been noted, or when determining site or structural responses including strong nonlinear behavior, phasing may play a more significant role. Therefore, it is recommended that the phase spectrum of appropriate recorded motions be closely maintained.

A special case occurs when developing time series for use in the standardized design of structures to be located across many sites. The design is therefore based on the use of a DRS. In this case, the tectonic environment and associated characteristic events have not been identified. Thus, the selection of time series needs to include sufficient power over a large frequency range of the DRS. One satisfactory approach is to develop synthetic acceleration time series using a randomly selected phase spectrum that has strong motion duration (duration from 5% to 75% Arias intensity) exceeding 36 s (McGuire et al. 2001). Alternatively, a suite of seed time series corresponding to the combinations of *M/D* and site characteristics for which the standardized design is intended to be sited may be used to develop modified recorded acceleration time series that correspond to the DRS.

Spectral accelerations at 5% damping should be computed at a minimum of 100 points per frequency decade, uniformly spaced over the log frequency scale over the frequency range of interest. This number of points eliminates concern with the accuracy of the computed spectrum and can easily be accommodated with current computer capability. If the target response spectrum is defined in the frequency range from 0.2 Hz to 25 Hz, the generated response spectrum should be compared with the target spectrum at each frequency in this range.

The objective is to generate a time series whose spectrum provides a mean-based fit to the target. Therefore, more points can fall below the target spectrum than has previously been required (USNRC 2014a). In USNRC (2014a), requirements are specified that allow no more than five points to fall below the target. Since the objective in this development is to generate motions whose spectra are not biased high or low with respect to the target, about as many points can fall below the target as can exceed the target. However, to prevent large frequency ranges where the spectrum falls below the target, no more than nine contiguous frequency points are allowed to fall below in one frequency band. This number of points corresponds to a frequency window ±10% centered on the frequency of interest, if the frequency points are selected as described here.

To ensure that no frequency gaps occur in the motion, the exceedance of the computed spectrum above the specified target spectrum is limited to no more than 30% of the target. The computed spectrum should still not fall below the target by more than 10%. If these general criteria are followed, the matching requirements to the 5% damped response spectrum should be adequate to ensure that no gaps in the PSD or Fourier amplitude spectrum will occur over a significant frequency range. There is then no special need to further evaluate the PSD of the ground motion. If the 5% damped response spectrum of the generated time series exceeds the target spectrum by more than 30%, a target PSD needs to be developed and the PSD of the time series needs to be generated and shown to be consistent with the target PSD at frequencies of interest in the analysis, as required in ASCE (2017b). Costantino and Houston (2008) and USNRC (2014a) provide guidance on developing target PSDs.

Directional correlation coefficients between pairs of acceleration records in a triplet to be used in site response and/or SSI analyses are typically required to be below some maximum value to ensure that the structure or structural element cannot be oriented in the analysis in such a manner as to minimize some important directional response quantity of interest. The commentary to Section 2.6.2 of ASCE 4-16 notes that the correlation of the three components of a ground motion set, two horizontal and one vertical, should be no greater than that calculated from recorded ground motions after scaling, with an upper limit of 0.3 and a mean value of 0.16, which are based on studies described in Hadjian (1981) and Huang et al. (2011, 2016). These recommendations are consistent with calculations made using the recorded database from McGuire et al. (2001), which showed the mean value of the correlation coefficient for all record sets (Western United States and CEUS) was 0.13 (NRC 2014b).

Analyses of structures, systems and components (SSCs) to ground motions and in-structure response spectra may require further refinement of the time step. If refinement is performed by interpolation in the time domain, errors may be generated that are propagated through the analysis. These errors may result in extraneous high-frequency content in the computed response. In general, ASCE 4 requires seismic response to be computed within 10% accuracy. Using this requirement, Mertz et al. (2016) indicated that the discretization error in a discretized time series meets the ASCE 4 criterion (<10%) for frequencies up to one-third of the Nyquist frequency. In a typical analysis using a 0.005 s time step, the Nyquist frequency is 100 Hz and the ASCE 4 requirement is met for frequencies up to 33 Hz when response spectra are calculated using the time integration method.

In discretizing existing acceleration time series, linear interpolation in the time domain generates spurious high-frequency responses and does not increase the high-frequency responses in a manner consistent with closed-form, steady-state responses. Interpolation of the record in Fourier (frequency) space retains the frequency content of the original input motion and results in an interpolated response that is consistent with the closed-form responses.

CHAPTER C3
EVALUATION OF SEISMIC DEMAND

C3.2 SEISMIC DEMAND DEVELOPMENT

Seismic demands as determined by the linear analysis methods may be used to establish the total demand acting on structural members. Chapters 5 and 8 in this standard provide the acceptance criteria for the load combinations and inelastic energy absorption factors F_{μ} for specific types of structures and limit states.

Demands on structural components of the seismic framing system are determined using the provisions of ASCE 4 (2017b). The numerical model of the seismic framing system should include all components that contribute mass and stiffness and materially affect the deformation of the seismic framing system. The numerical model of the framing system should include all sources of seismically reactive mass. A product of the analysis according to ASCE 4 is input motions to supported structures, systems, and components (SSCs) that might contribute mass but not stiffness to the seismic framing system.

C3.3 MODELING AND INPUT PARAMETERS

C3.3.1 Effective Stiffness of Reinforced Concrete and Steel-Plate Composite Wall Members. ASCE 4 (2017b), Section 5.5, provides information for probabilistic analysis and approaches for establishing structural variability.

C3.3.3 Damping Values for Structures, Systems, and Components. Table 3-1 reproduces information presented in Chapter 3 of ASCE 4 (2017b). Commentary is provided in ASCE 4 (2017b) and is not repeated here. Table 3-1 includes additional information on damping values for SC walls and HVAC.

Damping values for heating, ventilation, and air conditioning (HVAC) provided in Table 3-1 are from Morante (2006). This reference also provides values of damping for supported SSCs not identified in Table 3-1.

Table 3-1 includes information on SC walls that was not available when ASCE 4 (2017b) was being compiled and is included here for completeness. The values of damping assigned to SC walls are 3%, 5%, and 10% at Response Levels 1, 2, and 3, as defined in ASCE 4 (2017b) and Table 3-2. These values are based on Akiyama et al. (1989), Epackachi et al. (2015c), and Seo et al. (2016).

Demands on structural components of the seismic framing system are determined using the provisions of ASCE 4. The numerical model of the seismic framing system should include all components that contribute mass and stiffness and materially affect the deformation of the seismic framing system. The numerical model of the framing system should include all sources of seismically reactive mass. A product of the analysis according to ASCE 4 (2017b) is input motions to supported SSCs that might contribute mass but not stiffness to the seismic framing system.

The primary purpose of the data in Section 3.3.3 are to enable calculations of demand. Response levels should be established for each supported SSC using the demand-capacity procedures of ASCE 4, which require analysis of the supported SSC, a calculation of demand, and a calculation of the ratio of demand to capacity (D_e/C). Alternately, damping associated with Response Level 1 according to Table 3-1 may be used for calculation of demands on supported SSCs.

In general, when computing soil–structure interaction effects deterministically, D_e/C ratios for the best estimate case of soil properties may be used to determine response levels.

Nonlinear analysis of supported SSCs may be performed, but Response Level 1 values of damping should be used to avoid duplication of the hysteretic energy dissipation that is a product of nonlinear action.

CHAPTER C4
STRUCTURAL CAPACITY

C4.1 STRUCTURAL SYSTEMS

The acceptable lateral force-resisting systems for new nuclear facilities are parsed in Table 4-1 into bearing wall systems, building frame systems, and moment-resisting frame systems.

Many nuclear structures provide a confinement function, and the integrity of the exterior walls and roof framing must be preserved. These walls and roof slabs are generally constructed of reinforced concrete. The earthquake-damaged state of walls and roofs that provide a confinement function must be limited to narrow cracks that can serve as particulate filters, which limit allowable story drifts and dictate a minimum amount and maximum spacing of reinforcement.

The height limits on lateral force-resisting systems presented in ASCE 7 for commercial buildings are set aside in this standard because (a) few safety-related nuclear structures are tall, (b) the inelastic deformation demands on safety-related nuclear structures are small, and (c) there is no technical basis for the height limits of ASCE 7.

Many lateral force-resisting systems that are acceptable for commercial buildings in regions of low to moderate seismic hazard are unacceptable for new nuclear facilities and are not listed in Table 4-1. Ordinary and intermediate framing systems, which may be acceptable for commercial building construction according to ACI 318 (2014), are unacceptable for new nuclear facilities and accordingly and are not addressed in ACI 349 (2013b). Dual framing systems were identified as an acceptable lateral force-resisting system in ASCE 43-05 (2005). These systems have an essentially complete gravity load-carrying space frame and resistance to horizontal earthquake loading provided by either shear walls or braced frames and special moment-resisting frames. The requirements of ASCE 43-05 for dual framing systems effectively attached a weak moment frame system to a stiff, strong braced frame or shear wall. These requirements provided no significant relief for the design of the primary lateral force-resisting system and no incentive for its use. Accordingly, dual framing systems have been deleted as an acceptable lateral force-resisting system in this standard.

Nonductile lateral force-resisting systems are not acceptable for new nuclear facilities because inelastic deformation capacity will be required for either design basis or beyond design basis shaking.

C4.2 COMPONENT CAPACITIES

C4.2.1 General. New nuclear facilities are analyzed according to ASCE 4, and they are designed and detailed according to this standard and referenced materials standards. Structural systems are analyzed and the framing system is designed component by component.

The force capacity of a component of a nuclear structure is determined using a materials standard, such as one of those listed in the provisions for acceptable framing systems. The use of prescriptive details for ductile response included in the standard should provide adequate deformation capacity. First-principles calculations of deformation capacity should be performed if alternate ductile details are proposed; the calculations should be peer-reviewed.

The force capacity of a structural component for design is calculated by multiplying the ultimate strength of the component by a capacity (strength) reduction factor that is less than unity. Ultimate or nominal strength should be calculated using a consensus standard based on ultimate strength design, but where such a standard does not exist, they may be calculated using a standard based on allowable stress design and increasing the allowable strength by a factor that depends on material type and action.

C4.2.2 Reinforced Concrete. Provisions for the calculation of the force capacity of reinforced concrete (noncontainment) components for seismic design are given in ACI 349. Design strength is calculated by multiplying nominal strength by a strength reduction factor, ϕ. Prescriptive details for components to enable ductile response are provided in ACI 349.

An equation for the shear strength of low aspect ratio (shear-critical) walls that was based in part on the studies of flanged walls by Barda et al. (1977) was provided in ASCE 43-05. Recent studies (e.g., Gulec et al. 2008, 2009) have shown that the ASCE 43-05 equation (and that of Barda et al. 1977) does not conservatively predict peak shear strength, and so it has been deleted from this standard.

Provisions for the calculation of the force capacity of components of a reinforced concrete containment vessel are provided in ASME (2015a), Section III, Division 2.

C4.2.2.1 Shear Strength of Reinforced Concrete Members. Walls and slabs in nuclear safety-related structures are generally thicker than those in conventional building construction. The predictive equation for the one-way, out-of-plane shear strength of these elements is based on tests of shallow slender beams without shear reinforcement.

A detailed evaluation of the shear strength of reinforced concrete beams constructed without shear reinforcement, calculated in accordance with Equation (11-3) of ACI 318 (2008) shows clearly that the calculated shear strength is not sufficiently conservative to support the performance goals of this standard for all member geometries encountered in nuclear safety-related structures. The size (or depth) correction factor, κ, was developed to ensure that the ASCE 43 performance goals are met for the range of depths and reinforcement ratios commonly encountered in such structures.

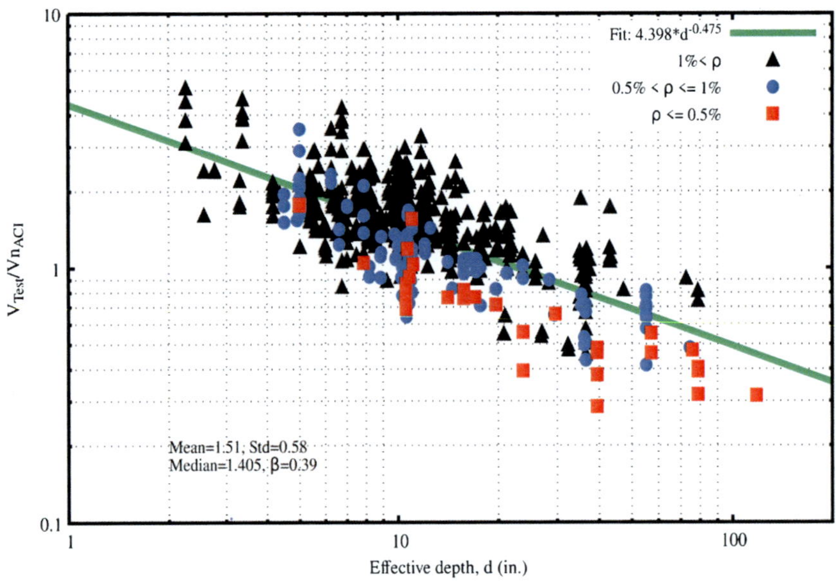

Figure C4-1. Ratio of test shear strength to nominal shear strength according to Equation (11-3) of ACI 349 (2013b).

Reineck et al. (2013) developed a database of 784 slender beams without stirrups. Figure C4-1 shows the ratio of test strength to that calculated using ACI 349 (2013b), Equation (11-3). The data show a clear trend of decreasing normalized strength with increasing effective depth, d, which is referred to as a *size effect*. Figure C4-1 also indicates that shear strength for members with a low (solid red squares) longitudinal reinforcement ratio, ρ, is smaller than for members with a high longitudinal reinforcement ratio (solid black triangles).

According to ACI 349 (2013b), the member design shear capacity is ϕV_n, where $\phi = 0.75$ and V_n is the nominal shear capacity. Failure of a slab or wall in out-of-plane shear is brittle and according to Section C1.5 of this standard, the required capacity to support the performance goal has a 98% probability of exceedance with an additional factor of conservatism of 1.33. Thus, $C_{98\%} \leq 1.33\phi V_n$, where $C_{98\%}$ is the 98% exceedance capacity.

The data in Figure C4-1 have a median value of 1.405 with a lognormal standard deviation of $\beta = 0.39$. Approximately 15% of the 784 test specimens have a shear strength less than V_n. Of the deeper members, namely, $d \geq 24$ in. (610 mm), 62 of 78 specimens (79%) have shear strength less than V_n. Clearly, the ACI capacity is not sufficiently conservative to support this standard's performance goal for one-way, out-of-plane shear in slabs and walls.

The data in Figure C4-1 have a trend line with a slope of approximately $1/\sqrt{d}$ where d is the effective depth in inches. The nominal shear capacity is factored by $2.8/\sqrt{d}$, and the resulting ratio of test strength to factored capacity is shown in Figure C4-2. Note that this factor reduces the dependency of capacity on depth. Furthermore, the variability of the data in Figure C4-2 is $\beta = 0.272$, which is substantially smaller than $\beta = 0.39$ for Equation (11-3) of ACI 349 (2013b), which indicates that $V_n(2.8/\sqrt{d})$ is a better representation of strength than V_n alone.

However, there is an unconservative bias for specimens with small longitudinal reinforcing ratios ρ, which may occur in nuclear applications where radiation shielding controls the thickness. This bias is evident in Figure C4-2 and a trend line through the data has a slope of approximately $(100\rho)^{0.25}$. Combining the depth and ρ corrections yields the factor $\kappa = 2.8(100\rho)^{0.25}/\sqrt{d}$ shown in Equation (4-1b). Multiplying the nominal shear capacity by κ yields the ratio of test strength to factored capacity, as shown in Figure C4-3. The data in Figure C4-3 are independent of both effective depth and reinforcing ratio. Furthermore, the variability of $V_{n\kappa}$ in Figure C4-3 is $\beta = 0.233$, which is smaller than the variability $\beta = 0.272$ for $V_n(2.8/\sqrt{d})$ and the variability $\beta = 0.39$ for Equation (11-3) of ACI 349 (2013b), which indicates that $V_{n\kappa}$ is a better fit to the actual member strength. Considering all of the data in Figure C4-3, 95% of the measured strengths are greater than $V_{n\kappa}$. Thus, $V_{n\kappa}$ alone does not provide sufficient conservatism to support the performance goals of this standard.

The in-service or expected concrete strength is typically greater than the minimum specified strength at the time of construction. Using a ratio of expected-to-specified strength of 1.2 with $\beta = 0.1$ yields $V_{n\kappa}$ with a 97.8% exceedance shear capacity. Considering the increase in compressive strength beyond 28 days, with an aging strength ratio of 1.2, $\beta = 0.14$, and noting that significant time passes between placing concrete and building commissioning in safety-related nuclear structures, yields $V_{n\kappa}$ with a 99% exceedance shear capacity. Thus, the factor of 2.8 in Equation (4-1b) is selected to ensure that the ASCE 43 performance goal is met for the range of member depths and reinforcement ratios commonly encountered in nuclear safety-related structures. The size reduction factor κ was developed using Equation (11-3) of ACI 349 (2013b) as the basis. Note that these increases in concrete strength may not be appropriate for nonnuclear reinforced concrete buildings.

As an interim measure, the factor is applied to all of the concrete shear strengths calculated in Section 11.2 of ACI 349 (2013b). The study did not address the size effect in precast members, and the engineer is cautioned to ensure that the design strength of precast members is consistent with the ASCE 43 required strength. A low volume of shear reinforcement is not effective in changing the shear behavior of reinforced concrete members. Thus, the demarcation between members with and without shear reinforcement is the minimum shear reinforcement in Section 11.4.6.3 of ACI 349 (2013b). This reduction affects the one-way, out-of-plane shear strength of walls, slabs, and footings that do not typically have shear reinforcement and may require the addition of shear reinforcement for some new designs. The data in Figure C4-1 are limited to one-way behavior, and it is

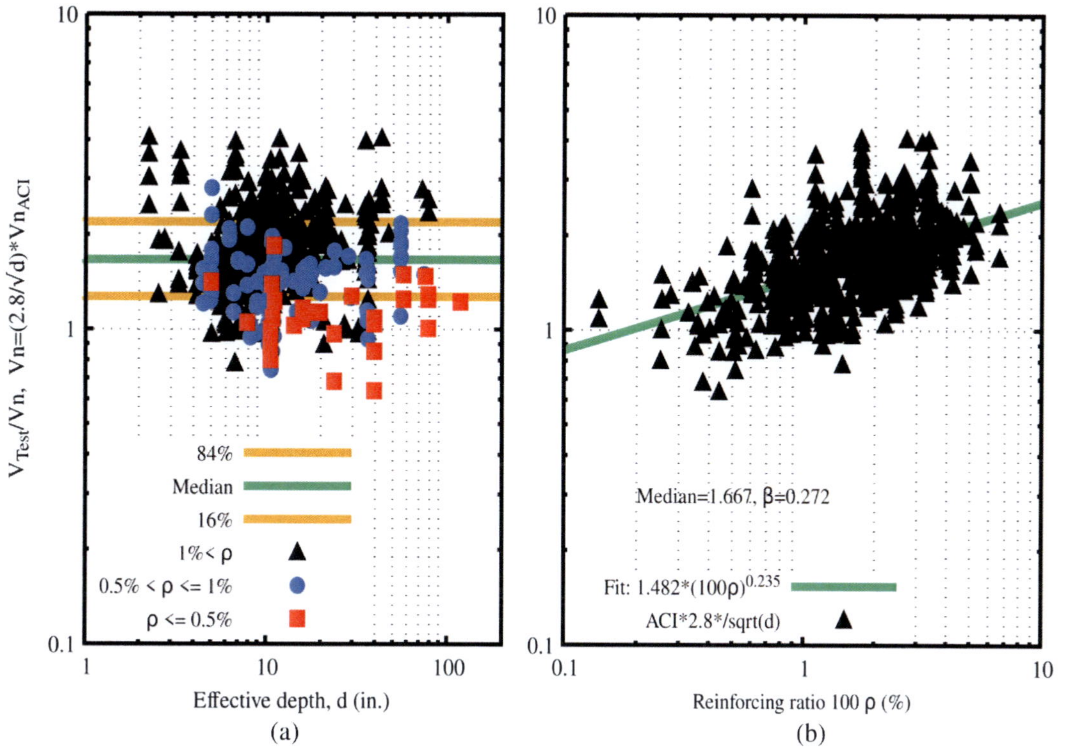

Figure C4-2. Ratio of test shear strength to $V_n(2.8/\sqrt{d})$.

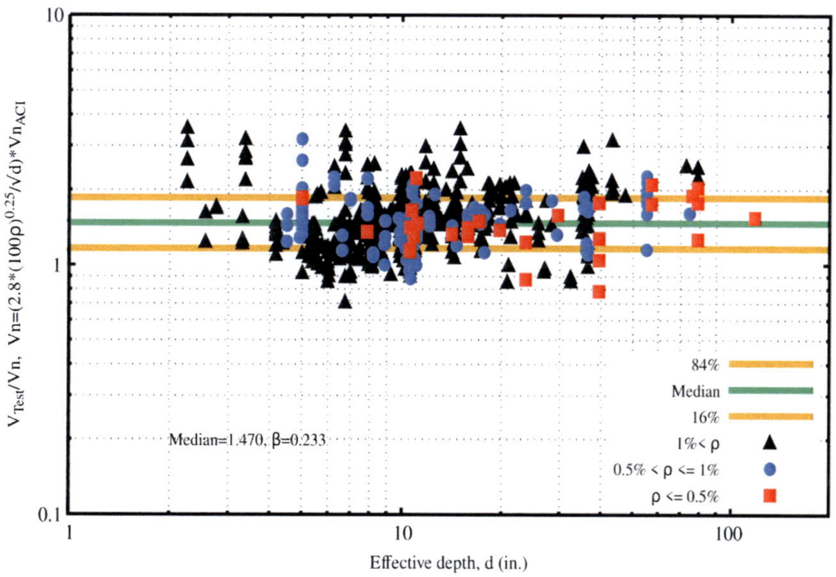

Figure C4-3. Ratio of test shear strength to $V_{n\kappa}$.

not known if the size effect also impacts punching shear. Importantly, the reduction may be unconservative (not sufficiently great) if there are coexisting forces, which in a wall may be simultaneous in-plane and out-of-plane shears and bending moments, and axial force. There are no data available at this time to support a further reduction from that proposed in Equation (4-1b).

C4.2.3 Structural Steel. AISC N690 (2018) enables the calculation of the force capacity of carbon steel (noncontainment) components for seismic design. AISC N690 (2018) does not address ductile response in the event of shaking more intense than design basis. AISC 341 (2016b) should be used to detail structural steel components for ductile response.

AISC N690 (2015) combined provisions for allowable stress design and ultimate strength design (load and resistance factor design). Design strength is calculated by either (a) multiplying a nominal strength by a strength reduction factor, or (b) increasing an allowable strength by 1.5, where the product of Ω and ϕ is 1.5 according to the commentary to AISC 360 (2016).

ASME (2015a) Section III, Division 1 NE, enables the calculation of the force capacity of components of a carbon steel containment vessel.

C4.2.4 Steel-Plate Composite. Supplement 1 to AISC N690 (2015) provides guidance on the design, detailing, and construction of steel-plate composite (SC) walls, with a focus on application to labyrinthine construction such as that found inside containment vessels of nuclear power plants. The seismic behavior of this relatively new framing system for nuclear structures (e.g., Varma et al. 2011, 2013; Epackachi et al. 2015b) is comparable to, or better than, reinforced concrete walls provided that the reinforcement ratio and faceplate slenderness ratio comply with the limits set forth in the supplement. The shear force capacity of SC wall piers (no boundary flanges) is less than that predicted by the supplement because pier behavior is not dominated by shear, but it can be estimated using Epackachi et al. (2015a).

C4.2.5 Reinforced Masonry. ACI 530 and ACI 530.1 (2013a) enable the calculation of the force capacity of reinforced masonry components for seismic design.

CHAPTER C5
LOAD COMBINATIONS AND ACCEPTANCE CRITERIA FOR STRUCTURES

The seismic load combinations in this standard combine a best estimate of the normal operating loads and the seismic load developed in Chapter 3. This standard has acceptance criteria for both linear and nonlinear analyses of structures. A force-based acceptance criterion is used for linear analyses, along with a check of the total elastic displacements. A displacement-based acceptance criterion is used for nonlinear analyses, with the provision that member capacities be limited to the product of the nominal material code capacity and the material code strength reduction factor, ϕ. Load combinations and acceptance criteria for systems and components are discussed in Chapter 8.

C5.1 LOAD COMBINATIONS

C5.1.1 General. The seismic load combinations are consistent with the seismic performance goals that form the basis of this standard and are intended to supersede seismic load combinations given in material codes (e.g., those of ACI and AISC), model building codes (IBC), and other standards, e.g., ASCE/SEI 7 (2017a). This standard presumes that load combinations for nonseismic loads are consistent with the material codes, model building codes, and other standards.

C5.1.2 Seismic Load Combinations

C5.1.2.1 Seismic Load Combinations for Strength-Based Acceptance Criteria. The seismic demand for elastic analyses is reduced by an inelastic force reduction factor, F_μ, to account for nonlinear energy absorption.

Elements That Experience Combined Response. As illustrated by Equations (5-1a) and (5-1b), the component inelastic energy absorption factor, F_μ, is allowed only for specific response modes. Many elements have combined response modes. An example is a column subject to combined axial force and bending. In this case, the F_μ factor is only applied to moment, generally in the following form:

$$\frac{P_{NS}+P_S}{\phi Pn}+\frac{M_{NS}+\frac{M_S}{F_\mu}}{\phi Mn}\leq 1.0 \qquad \text{(C5-1)}$$

where

P_{NS} = Nonseismic concurrent axial force demand,
M_{NS} = Nonseismic concurrent moment demand,
P_S = Seismic axial force demand,
M_S = Seismic axial moment demand, and
ϕPn and ϕMn = Code capacities consistent with the acceptance criteria in Section 5.2.

C5.1.2.2 Seismic Load Combinations for Deformation-Based Acceptance Criteria. Nonlinear lateral displacements for an elastic analysis are approximated by $\mu\Delta_{\text{elastic}}$. Calculating the elastic lateral displacements with Equation (5-1a) and assuming that the force reduction factor, F_μ, is approximated by μ, then the nonlinear displacement is

$$\Delta_{NL}=\mu\times\Delta_{\text{elastic}}=\frac{\mu\times 1.0\times\frac{D_S}{F_\mu}}{\text{Stiffness}}\cong\frac{1.0\times D_S}{\text{Stiffness}} \qquad \text{(C5-2)}$$

which is similar to the displacement calculated with Equation (5-2). Thus, Equation (5-2) is used to estimate the nonlinear displacements for an elastic analysis. Note that F_μ is approximated by μ in the displacement portion of the spectra and is less than μ in the velocity and acceleration portions of the spectra. Thus, Δ_{NL} is larger than the elastic displacement with 1.0 D_S at medium to high frequencies. However, displacements in this frequency range are typically small and do not control the design. Thus, this nonconservatism is acceptable.

C5.1.3 Inelastic Energy Absorption Factor. The engineering community has known for some time that ductile structures are better at resisting seismic loads than are brittle structures. Building codes for conventional structures recognize the benefits of ductile construction and allow for design using a reduced seismic base shear to reflect the beneficial effects of ductile behavior. This reduced base shear is calculated using force reduction factors. However, building codes for conventional structures are primarily concerned with preventing the loss of human life and allow significant structural damage. This level of damage may not be appropriate for facilities containing nuclear materials. For example, it may be desirable to limit the seismic damage in a nuclear confinement structure to hairline cracks, whereas a non-safety-related structure could suffer appreciable damage as long as it did not collapse on a safety-related structure. Thus, an appropriate set of design rules for a nuclear structure would allow for various force reduction factors, depending on the amount of damage that is acceptable. This acceptable damage level defines the limit state.

Overstrength. A force reduction factor, R, that accounts for both the overstrength and the inelastic energy absorption of the structural system is used in the model building codes for conventional structures. Overstrength in a building system can be quantified using a pushover analysis or can be estimated from existing analyses of similar structures. Pushover analyses are not usually performed as part of the design process, and the use of existing analyses to estimate overstrength is problematic for nuclear facilities. Thus, the overstrength factor is conservatively omitted from the force reduction factors in Table 5-1. The overstrength factor in typical nonnuclear commercial building frames ranges from 2 to 3.5 and represents about one-half of the

R factor used in building codes (ATC 1974). An overstrength for a particular structure, which is quantified by a pushover analysis, may be included in the development of a facility-specific force reduction factor, F_μ.

Development of Inelastic Absorption Factors. Inelastic energy absorption factors have been extensively studied. The element ductility, amount of strain hardening, shape of the hysteresis loops, building frequency, supporting soil type, and the influence of higher modes (single degree of freedom [SDOF] versus multiple degrees of freedom [MDOF]) affect the inelastic energy absorption factor.

The inelastic energy absorption factors in this standard are developed by

1. Selecting element ductilities, $\mu_{element}$, that are correlated to desired performance or limit state;
2. Relating frequency dependence, secondary effects of strain hardening, hysteresis loop shape, and soil type to the force reduction factor for an SDOF system;
3. Developing relationships between the ductility of MDOF and SDOF systems; and
4. Combining items 1, 2, and 3 to yield inelastic energy absorption force reduction factors for design.

These four steps are described in the subsections below and are shown schematically in Figure C5-1.

The final calculated inelastic absorption factor, F_μ, is referred to as a system-level inelastic absorption factor. The term system is used because the factor is developed from the component-level inelastic absorption factor, $F_{\mu C}$, and modified to characterize system-level effects. However, the F_μ is not representative of system response in the same sense as the response modification factor, *R* value, of the International Building Code (ICC 2015).

Existing Structures with Noncompliant Detailing. This standard requires that new structures, systems, and components (SSCs) be designed to the ductile detailing requirements of Chapter 6. Existing structures may contain elements that have noncompliant detailing. The following discussion is limited to the evaluation of existing structures.

A project-specific system energy absorption factor, F_μ, and deformation limits for noncompliant detailing may be developed using the methodology outlined in the section titled "Development of Inelastic Energy Absorption Factors." Section C1.5.1.2.3 states that an inelastic factor established at a 5% nonexceedance probability is consistent with the performance goals of this standard.

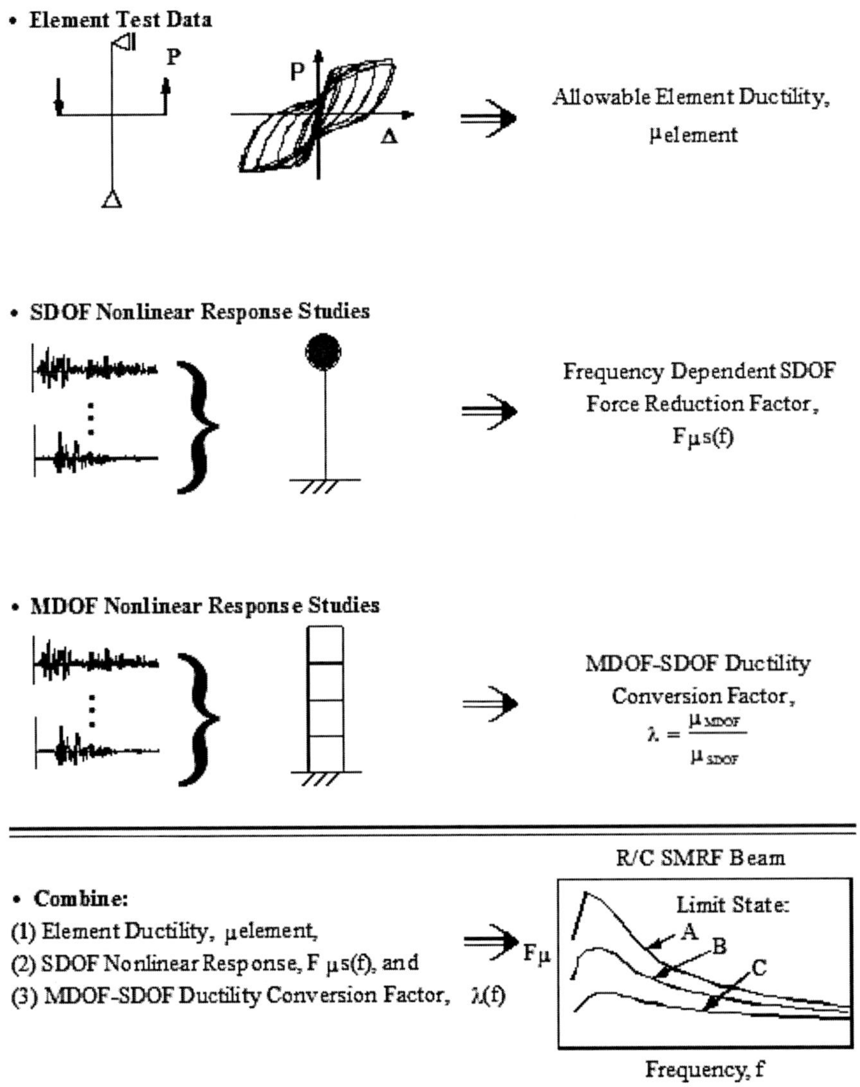

Figure C5-1. F_μ **development.**

ASCE/SEI 41 (2013) contains a rich collection of element deformation limits for various levels of nonconforming detailing in tables titled "Modeling Parameters and Acceptance Criteria for Nonlinear Procedures," which may be used as input for the development of a project-specific system energy absorption factor, F_{μ}. A predecessor to ASCE/SEI 41 (2013), FEMA 273 (FEMA 1997) was used as input to the development of the inelastic factors in this standard (Mertz 2001).

The ASCE/SEI 41 (2013) data are developed for different limit states from those used in this standard (Section 2.3 of ASCE 2013).

- The ASCE/SEI 41 (2013) immediate occupancy limit state corresponds to the deformation at which permanent visible damage occurs but generally not greater than 0.67 times the life-safety limit state. The immediate occupancy limit state is roughly equivalent to Limit State C of this standard.
- The ASCE/SEI 41 (2013) life-safety limit state for primary members is 0.75 times the deformation corresponding to a significant loss of load capacity. Assuming a coefficient of variation of 0.25, this value corresponds to roughly a median minus one sigma level (sigma denotes standard deviation). The life-safety deformation limit is roughly equivalent to Limit State A of this standard.
- ASCE/SEI 41 (2013) has a collapse-prevention limit state that considers the deformation capacity after significant loss of lateral load capacity. This standard does not consider deformations beyond the significant loss of lateral load capacity.
- The deformation limits for Limit State B are the average of the deformation limits for Limit States A and C.

Experimentally determined capacities in ASCE/SEI 41 (2013) are established at mean minus one standard deviation, which is less conservative than the code-level capacities required by this standard.

ASCE/SEI 41 (2013) uses *m*-factors to account for inelastic energy dissipation. The *m*-factors are based on 75% of the deformation limits referenced above, divided by the mean yield deformation.

Consideration of Response Level 3 Damping with F_{μ}. The nonlinear cyclic behavior of low-rise shear walls is dominated by pinched hysteresis loops. Pinched hysteresis loops dissipate less energy than full hysteresis loops encountered in special moment frames. Thus, shear walls are used in Bozard (2002) to study the interaction between deformation limits, F_{μ}, damping, soil–structure interaction (SSI) response and weak or soft story effects. Bozard (2002) indicates that the F_{μ} factors in this standard are conservative for structures with and without SSI, where the SSI response may include significant radiation damping. The study also indicates that concentrating the inelastic deformation in the shear wall, combined with SSI flexibility, does not require a weak or soft story reduction factor. In addition, the study demonstrates that the F_{μ} factors are generally conservative when combined with Response Level 3 damping and SSI with significant radiation damping. Thus, it is the judgment of the DANS committee that the F_{μ} factors in this standard may be combined with Response Level 3 damping in accordance with Section 3.3.3 of this standard and the F_{μ} factors may be used in structures with or without SSI effects.

C5.1.3.1 Component Inelastic Energy Absorption Factor. Acceptable damage for each different limit state (LS) is defined for each type of element. Four limit states are used in this standard that range from moderate damage in LS A to essentially elastic behavior in LS D. Intermediate limit states consist of LS C, which corresponds to Immediate Occupancy, and LS B, which is defined as halfway between LS A and LS C. Story displacements or element rotations associated with each limit state are selected based on the average element behavior, as shown in Figure C5-2. The summary of element data in ASCE (2013) and Duffey et al. (1994) form the basis for the element ductilities and allowable deformations used in this standard. The element is conservatively assumed to have an elastoplastic behavior with a yield point at the nominal code capacity and an allowable inelastic deformation corresponding to the appropriate limit state displacement. Element ductilities for each limit state are derived on this assumption.

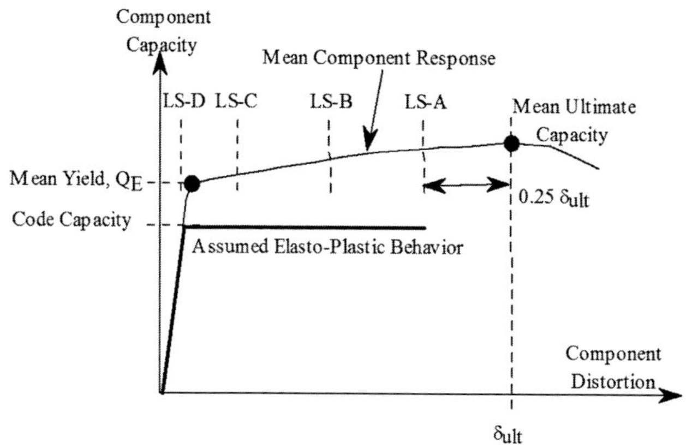

Figure C5-2. Typical load-deformation curve and limit states.

The inelastic response of different single degree of freedom (SDOF) systems to various ground motions have been extensively studied (Newmark and Hall 1987, Nassar and Krawinkler 1991, Miranda and Bertero 1994, Kennedy et al. 1984). Flexural structural systems are represented by hysteresis models that have full stable hysteresis loops, and structural systems dominated by shear deformation or diagonal brace buckling are represented by pinched hysteresis models. Relationships developed by several researchers are examined, and force reduction factors for an SDOF system, $F_{\mu\ SDOF}$, are determined considering structural system type, system frequency, and element ductility. These force reduction factors have a conservative bias when compared to literature relationships (Mertz 2001).

The resulting frequency-dependent F_{μ} values are functions of limit state and natural frequency, as shown in Figure C5-1. Conservatively, frequency-independent F_{μ} factors for each type of framing system are selected based on building frequencies representative of that typically found in facilities containing nuclear materials. These frequency-independent F_{μ} factors are given in Table 5-1. The use of frequency-dependent F_{μ} factors, such as those given in Mertz (2001) is permissible.

SDOF models neglect the contribution of higher modes and the relationship between component and global ductility. For some structures, these factors may be important, especially at lower frequencies. Relationships between the response of MDOF systems and SDOF systems are presented in the literature (Nassar and Krawinkler 1991; Osteraas and Krawinkler 1990; Seneviratna and Krawinkler 1993, 1997; FEMA 2009). The allowable element ductility, modified to account for MDOF effects, is used to develop frequency-dependent inelastic energy absorption factor for elements used in each of the building types. The relationship between SDOF and MDOF F_{μ} for structures that have a soft or weak story (Mertz 2001) is not included in the $F_{\mu C}$ in Table 5-1 and is applied separately in this standard.

Shear Walls. The F_μ given for shear walls is for in-plane bending moments and shears. Out-of-plane, a shear wall can be considered as a beam with a low axial load. Shear walls governed by out-of-plane shear are brittle unless special out-of-plane shear reinforcing has been installed. Brittle elements have $F_\mu = 1.0$.

Anchorage. To justify an inelastic factor F_μ greater than 1.0 for any specific anchorage situation, it is necessary to specify the nonlinear distortion capacity of the anchorage corresponding to the 95% probability of exceedance for the specified limit state. Based on the specified anchorage distortion capacity, the method presented in the following section, "Alternate Method of Estimating the Inelastic Energy Absorption Factor," may be used to determine the system inelastic energy absorption factor F_μ for the anchored component, which may then be used as F_μ for the anchorage.

However, in most cases the 95% probability of exceedance distortion capacity of the anchorage is insufficient when compared to the anchored component displacement to justify an F_μ significantly larger than 1.0.

Alternate Method of Estimating the Inelastic Energy Absorption Factor. The method described here may be used to estimate the inelastic energy absorption factor. In general, the analyst would first make an estimate of a permissible inelastic distortion corresponding to about a 5% failure probability level. Once a permissible distortion has been selected, the inelastic energy absorption factor F_μ may be determined from nonlinear analysis of an appropriate model of the structure using multiple realistic input time histories.

First, the input time history is scaled to a level at which the elastically computed demand is equal to the yield (or ultimate) capacity. Then the input is further scaled until the distortion resulting from a time history nonlinear analysis reaches a permissible value. The inelastic energy absorption factor F_μ is equal to this additional scaling factor.

Alternately, a simplified nonlinear analysis procedure may be used. First, the analyst estimates the nonlinear deformed shape of the structure corresponding to the maximum permissible distortion that is reached in the story with the highest value of the ratio of the demand to the capacity.

Then the system ductility μ is estimated from

$$\mu = \frac{\Sigma W_i \delta_{Ti}}{\Sigma W_i \delta_{ei}} \tag{C5-3}$$

where W_i is the inertial weight applied at each story of the structure, δ_{Ti} is the total displacement relative to the base of each story corresponding to the permissible total distortion occurring in the critical story, and δ_{ei} is the elastic displacement relative to the base corresponding to a unit value of the ratio of the elastic demand to the capacity for the critical story. See Kennedy et al. (1988) for additional details.

For a single story, Equation (C5-3) simplifies down to a story ductility, μ_s, of

$$\mu_s = \frac{\delta_T}{\delta_y} \tag{C5-4}$$

where δ_T is the total permissible story displacement and δ_y is the yield displacement. However, for multistory structures, μ from Equation (C5-3) is always less than μ_s from Equation (C5-4), except in the unlikely case that the nonlinear distortions are spread throughout the structure.

The following equation relates μ to μ_s:

$$\mu = 1 + F_k(\mu_s - 1) \tag{C5-5}$$

where F_k is a reduction factor used to convert a story ductility to give a system ductility estimate. For a well-designed structure in which the ratio of demand to capacity does not differ by more than a factor of about 1.3 over the structure height, F_k typically ranges from 0.5 to 0.75. In these cases, Equation (C5-5) may be used with a conservative estimated F_k of 0.5 in lieu of Equation (C5-3) or nonlinear time history analyses.

Once the permissible system ductility μ has been established, many approaches can be used for estimating F_μ. For broad, smooth input response spectra and moment-frame structures with essentially full elastoplastic nonlinear hysteretic loops, either the Newmark and Hall (1978) or the Riddell and Newmark (1979) approach is commonly used. However, for concrete shear wall structures or braced frames that have severely pinched hysteretic loops, Kennedy et al. (1984) have shown that these approaches are likely to be slightly nonconservative. For such structures, the approach of Kennedy et al. (1984) is preferred.

C5.1.3.2 Weak or Soft Story Inelastic Energy Absorption Factor. **Weak Story Effect.** The F_μ modification for weak story effects is based on Equation (5-3) (Mertz 2001). This formulation is based on the assumption that all of the inelastic deformation is concentrated into a single weak story of an *n*-story frame. Equation (5-3) was developed for displacement-controlled structures where $F_\mu = \mu$ and is conservative for velocity and acceleration-controlled structures. Numerical studies have also demonstrated that Equation (5-3) is conservative for reasonable variations in mass and stiffness.

C5.1.3.3 High-Frequency Inelastic Energy Absorption Factor. **Effective Natural Frequency**. Force reduction factors for rigid frames are highly dependent on frequency, as discussed in Section C5.1.3.1, and Mertz (2001) developed relationships for these force reduction factors at different frequencies. The inelastic energy absorption factors in Table 5-1 are based on the assumption that rigid frames have natural frequencies less than about 3 Hz, which is typically at or below the frequency corresponding to the peak spectral acceleration.

For frequencies corresponding to a range of decreasing spectral accelerations, the softening of the frame caused by inelastic action can shift the frame's effective natural frequency to a region of higher spectral acceleration, as shown in Figure C5-3. In Mertz (2001), this effect is included in the frequency-dependent F_μ. In this standard, a simplified approach is used to account for this effect by using the spectral acceleration of the effective natural frequency, f_e.

Given the idealized load deflection diagram in Figure C5-4, the elastic stiffness is K_e, and the secant stiffness of a nonlinear frame with ductility μ can be expressed as

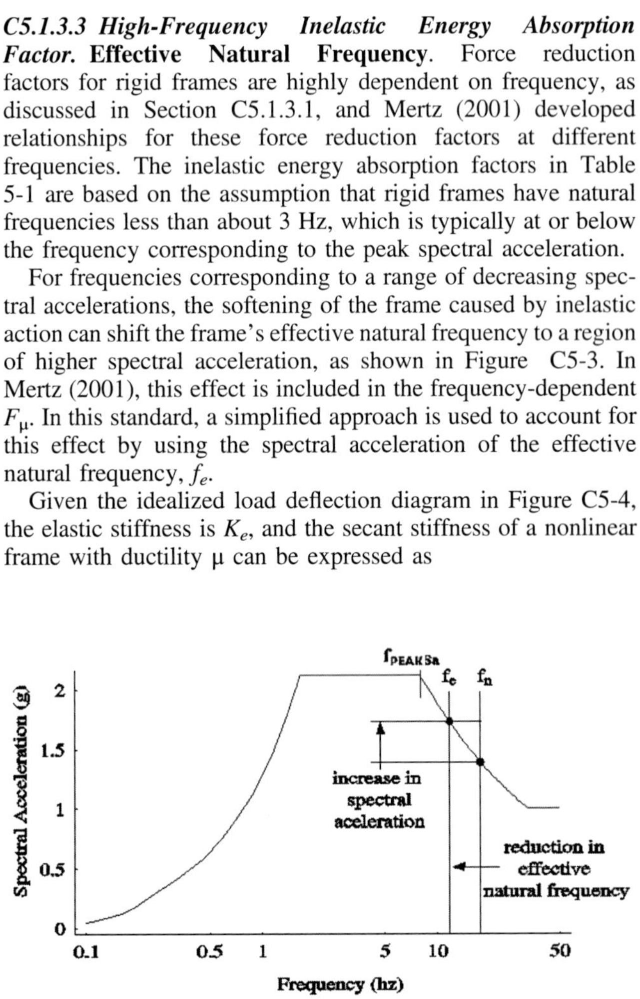

Figure C5-3. Idealized design response spectra.

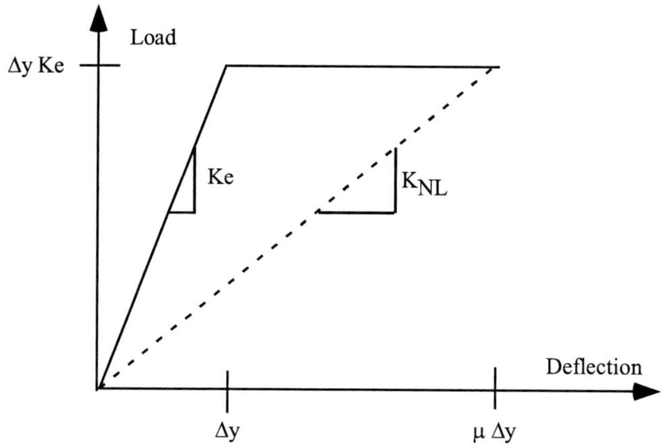

Figure C5-4. Idealized load deflection curve.

$$K_{NL} = \frac{K_e \, \Delta y}{\mu \, \Delta y} = \frac{K_e}{\mu} \quad \text{(C5-6)}$$

Assume that the effective frequency, f_e, is based on the nonlinear secant stiffness, K_{NL}. This assumption leads to a conservative lower bound effective frequency. The ratio of the effective to elastic natural frequency, f_n, is

$$\frac{f_e}{f_n} = \frac{\sqrt{\frac{K_{NL}}{M}}}{\sqrt{\frac{K_e}{M}}} = \sqrt{\frac{1}{\mu}} \quad \text{(C5-7)}$$

Because F_μ values for rigid frames are based on about 3 Hz, which is typically in the constant velocity range, then $F_\mu \approx \sqrt{2\mu - 1}$, and a conservative estimate of the nonlinear frequency is

$$f_e = \sqrt{\frac{2}{F_\mu^2 + 1}} f_n \quad \text{(C5-8)}$$

The minimum value of the effective frequency is limited to $f_{\text{PEAK } SA}$ to force the use of the peak spectral acceleration for structures that soften through the peak of the spectra.

This discussion is based on rigid frames. Structures with pinched hysteresis loops, such as low-rise shear walls and braced frames, have F_μ values that do not vary significantly between 2.0 and 8.5 Hz (Kennedy et al. 1984). Thus, the application of Equation (5-4) may be overly conservative for shear wall structures and can be relaxed with engineering justification.

C5.1.3.4 Adjustments to the Inelastic Energy Absorption Factor for Ratcheting. Ratcheting can occur when external forces act in the same direction as the seismic load. If these forces are sufficiently large, they have the potential to prevent the structure from unloading and to cause displacements that are significantly larger in one direction than the other. Elements with a potential for ratcheting include basement walls with at-rest lateral soil loads subject to dynamic lateral soil loads and slabs or beams with gravity loads subject to vertical seismic motion.

The factor α_{F_μ} was developed to convert inelastic deformations that have an 84th percentile probability of nonexceedance without ratcheting effects to those for systems that have ratcheting potential, as indicated in Eggers and Baig (2015). The analysis considered a suite of 30 recorded time history motions for an elastoplastic hysteresis and a postyield hardening of 1%, similar to the methodology used in Mertz (2001).

C5.2 ACCEPTANCE CRITERIA

This standard uses force-based acceptance criteria for linear analysis and displacement-based acceptance criteria for nonlinear analysis. Both linear and nonlinear analyses are evaluated for load combinations representing the expected dead and live loads combined with the mean seismic load.

For linear analyses, the force-based acceptance criterion is that the total demand in each element is less than or equal to the element's capacity. The seismic portion of the total element demand is reduced by an inelastic energy absorption factor, F_μ. Drift limits in Table 5-2 are specified for the total displacement. These nonlinear displacements are estimated with an elastic analysis using the load combination in Equation (5-2).

Both story drift limits and individual element drift limits are specified for nonlinear analyses. Most nonlinear analyses focus on yielding in selected elements and consider the remaining elements to be elastic. The elastic elements in a nonlinear analysis may be evaluated by the nonlinear criteria. Alternately, the elastic elements in a nonlinear analysis may be evaluated using the strength-based acceptance criteria with $F_\mu = 1.0$. Elastic elements, in a nonlinear analysis requiring a larger F_μ, should be modeled as nonlinear elements.

C5.2.3 Deformation-Based Acceptance Criteria

C5.2.3.1 Allowable Drift Limits for Structural Systems. The limits on allowable drift (Table 5-2) and on allowable rotation (Table 5-3) represent the maximum allowable deformations for a given limit state, as shown schematically in Figure C5-4. These deformation limits were derived from ASCE (2013) and Duffey et al. (1994) as part of the first step in F_μ development (Mertz 2001).

Table 5-2 contains story drifts, which are applied to both linear and nonlinear analyses. These drifts represent both elastic and inelastic deformations. The drift limits for braced frames and shear walls are the total deflections from the literature. The drift limits for moment frames include the nonlinear hinge rotation and an assumed elastic drift ratio of about 1/200. The drift limit for LS D consists of the smaller of the 1/200 elastic drift ratio or the LS C drift ratio.

Table 5-3 contains nonlinear hinge rotations for rigid frame structures. The nonlinear deformations of frame elements are assumed to be lumped into nonlinear hinges, and limits on the nonlinear hinge rotation are specified. The summation of nonlinear hinge rotations and the elastic deformation of the element between hinges is taken as the total element deformation.

CHAPTER C6
DUCTILE DETAILING REQUIREMENTS

The goal of this chapter is to invoke ductile details in components to enable stable nonlinear behavior of structural systems in design basis (DB) and/or beyond design basis earthquake (BDBE) shaking. Ductile details enable large inelastic deformation capacity in cross sections of components. Connections of components are generally designed to be stronger than the weakest connected component to force inelastic deformation into appropriately detailed cross sections.

The inelastic energy absorption (ductility) factors given in Chapter 5 imply that inelastic deformation is permissible when structures are subjected to extreme earthquake loads. Inelastic deformation is limited as a function of the intended limit state. Inelastic deformation of a component corresponds to damage, which may require significant repair.

This chapter addresses (a) materials, connections, and anchorages, which provide ductile response at the component level, and (b) plan and vertical irregularities, distribution of strength over the height of a structure, redundancy, and seismic interaction, which enable ductile response at the component level.

A regular structure with decreasing strength and stiffness over its height can engage a significant number of components with inelastic deformations approaching the deformation limits in Chapter 5. Contrarily, an irregular structure and/or a structure with a different distribution of strength and stiffness may engage only a small number of components in the energy dissipation system, with a much smaller energy dissipation capacity per unit drift than in regular structures. Minimizing irregularities and maintaining appropriate distributions of stiffness and strength are important for achieving robust, ductile seismic response.

Experience gained through evaluating and upgrading high-hazard safety-related nuclear structures for seismic effects has provided the impetus for some of the detailing requirements introduced in Chapter 6.

C6.1 STEEL STRUCTURES

AISC N690 (2018) provides procedures to design and detail steel structures for design basis loadings but does not provide guidance for detailing components for ductile response to loadings in excess of design basis.

Acceptable steel seismic framing systems are listed in Chapter 4 and Section 6.1. The detailing requirements of ANSI/AISC 341 (2016c) are invoked to ensure ductile response required to support the inelastic force reduction factors, F_{μ}, in this standard. Prequalified special moment frame (SMF) joints should be used, according to ANSI/AISC 358 (2016b).

When applying the detailing requirements of ANSI/AISC 341 (2016c), a system overstrength factor of unity is consistent with the development and intent of the component-specific energy absorption factors, F_{μ}, as presented in Chapter 5. The provisions of ANSI/AISC 341 are applicable to all nuclear structures irrespective of the seismic design category (SDC) referenced by AISC 341.

C6.1.4 Collectors. Section 6.1.4 presents provisions for that are important for providing a robust, complete seismic load path.

In structures with composite metal decks, headed studs in addition to those provided for composite behavior should be provided to transfer seismic inertial forces into collector beams.

If steel beams are used to collect and transfer seismic forces to the seismic load-resisting system, their connections should be designed for collector axial loads in addition to gravity shear loads. The engineer is cautioned against using long-slotted or short-slotted holes in bearing connections resisting collector forces. Double web framing angles in connector elements have complex load paths, including angle bending and prying actions, and these complexities should be addressed explicitly.

If unfilled metal deck is used to laterally support collector elements, combined in-plane and out-of-plane deck loads should be considered in addition to stability loads. The finite stiffness of the unfilled metal deck to collector connection should be addressed explicitly in any evaluation.

C6.1.5 Nearly Rigid Platforms and Supports. Platforms and supports that are very stiff are often used to support equipment and/or systems with minimal amplification of seismic input motion. Since rigidity is the primary design objective, these structures typically have low stresses. In these cases, it is neither warranted nor practical to size connections to develop the capacity (strength) of the weakest connected component. Although Section C1.5 recognizes that low-ductility failure modes typically have a factor of conservativism of 1.33, which meets the performance goal with $F_{\mu} = 1.0$, the committee determined that the seismic demand should be increased by a factor of 1.5 if ductile detailing is not implemented.

C6.2 REINFORCED CONCRETE

C6.2.1 General. The requirement to resist shear corresponding to 125% of the nominal moment capacity of the member is similar to the requirements of Sections 21.5.4 and 21.6.5 of ACI 349 (2013b). In this standard, this requirement is extended to address out-of-plane bending of walls.

In some structural walls, in-plane shear and bending dictate their design and detailing. Consequently, the out-of-plane flexural capacity may be significantly greater than that required for out-of-plane loads, and it may not be practical to design for out-of-plane shear corresponding to 125% of the nominal out-of-plane moment capacity of such walls. Although Section C1.5

recognizes that low-ductility failure modes typically have factor of conservatism of 1.33, which meets the performance goal with F_{μ}=1.0, the committee determined that that the seismic demand should be increased by a factor of 1.5 if ductile detailing is not implemented. This provision is not applicable for beams and columns in moment frames designed according to Sections 21.5 and 21.6 of ACI 349-13 (ACI 2013b).

If nonlinear analysis of a mathematical model including walls assigned an out-of-plane flexural strength of 125% of their nominal out-of-plane capacity is performed to generate the out-of-plane shear demand, the increase in seismic demand by a factor of 1.5 is not required.

C6.2.2 Slab-Wall Moment Frame Systems. A slab-wall moment frame is a moment-resisting frame, composed of both walls and slabs, which resists seismic lateral loading by out-of-plane bending, as shown in Figure C6-1. Perpendicular loads are resisted by in-plane shear in the slabs and shear walls. Slab-wall moment frames are typically found in nuclear materials processing facilities, where the slab and wall thickness are dictated by shielding requirements and not by load effects.

Below-grade walls subject to lateral soil loads with a horizontal wall span, measured between perpendicular shear walls, greater than twice the slab span, should also be considered as slab-wall moment frames, unless it can be demonstrated that the lateral loads can be transferred to the perpendicular walls through diaphragm action.

Slab-wall moment frames typically have lower reinforcement ratios, lower shear stresses, and lower axial stresses than special moment-resisting frames (SMRFs), which suggests that slab-wall moment frames are more ductile than SMRFs. However, slab-wall moment frames do not usually have transverse reinforcement along their entire length, and the splices in slab-wall frames may be located in potential hinging regions. In the absence of test data, slab-wall moment frames are assigned a limited state (LS) rotational capacity of 0.01 radians.

In slab-wall frame systems, the wall-slab connection, or joints, is subjected to high shear forces, especially when the slab or the wall or both form hinges adjacent to the joint. To ensure structural integrity, concrete in the joint should be confined. Typical confinement details are shown in Figure C6-2.

Section 6.2.2 uses the strut-and-tie method to analyze joint strength. One method of reinforcing a joint with the strut-and-tie method is shown in Figures C6-2a and C6-2b, where diagonal crossties are used to control cracking for both opening and closing moments. The crossties should engage, at least, every other vertical rebar with a maximum bar spacing of 12 in. (305 mm) on center. The crossties should be anchored with seismic hooks. Alternately, horizontal ties, as shown in Figures C6-2c and d may be used to provide joint confinement. Depending on the joint configuration, horizontal, vertical, or both horizontal and vertical ties may be required. The ties should engage, at least, every other perpendicular rebar with a maximum bar spacing of 12 in. (305 mm) on center. These ties may consist of (1) hoops or (2) crossties with seismic hooks. Hoops need not overlap.

The interaction between joint shear caused by frame action and joint shear caused by shear-wall action (in-plane shear) may be neglected.

C6.2.3 Requirements for Members Not Proportioned to Resist Forces Induced by Earthquake Motions. Chapter 21 of ACI 349-13 (2013b) provides ductile detailing requirements for special moment resisting frames, shear walls and coupling beams, and diaphragms. Chapter 21 does not provide sufficiently explicit requirements for beams, columns, and/or flat slabs in shear wall buildings that resist gravity loads and vertical seismic loads. The requirements of Section 21.3 of ACI 318-08 (ACI 2008) say that "Members not designated as part of the seismic-force-resisting system" are judged to provide adequate deformation capacity to beams, slabs, and columns of gravity frames subject to vertical seismic loads and horizontal deformations that are limited by the primary lateral load-resisting system. The provisions of ACI 318-08 (2008) are invoked because the provisions of ACI 349-13 (2013b) are based on ACI 318-08.

Columns support flat slabs and flat plates, but their lateral (horizontal shear) strength should not be included in the seismic load-resisting system. Such columns and slabs must be designed and detailed to resist earthquake-induced forces and deformations.

C6.2.4 Collectors. Collector forces may be determined from an adequate finite element analysis. Collector forces may appear in such analysis as a nonuniform membrane tension and may be distributed over a significant width. Results may be sensitive to the choice of mesh. Collector forces may also be obscured by diaphragm bending under orthogonal lateral loading.

Alternately, collectors may be proportioned to transmit an axial force directly to a shear wall equal to the tributary width of uniform diaphragm shear, consistent with the ACI equations for nominal shear capacity of a diaphragm. The collector is designed for a tensile force equal to the uniform diaphragm shear times a tributary width. Figure C6-3 presents an example calculation of collector forces.

Collector forces should be resisted by reinforcement in addition to diaphragm or slab reinforcement, even if excess capacity is provided in them. The additional steel prevents the formation of a weak link in the building load path under beyond-design basis loads.

The collector reinforcement should be anchored in the shear wall a sufficient distance to avoid exceeding the wall shear capacity in the load transfer region.

Collectors with large axial stresses and long lengths may have significant lateral deformations; these deformations may damage other structural elements. These deformations should be addressed in the analysis, design, and detailing of the structure. Axial stresses in the collectors may need to be minimized to preclude unacceptable damage to other structural elements. Moehle et al. (2010) provide an example that demonstrates the importance of considering collector deformation.

C6.2.5 Joints in Floor Slabs. Control or expansion joints located in the seismic load path serve as planes of weakness

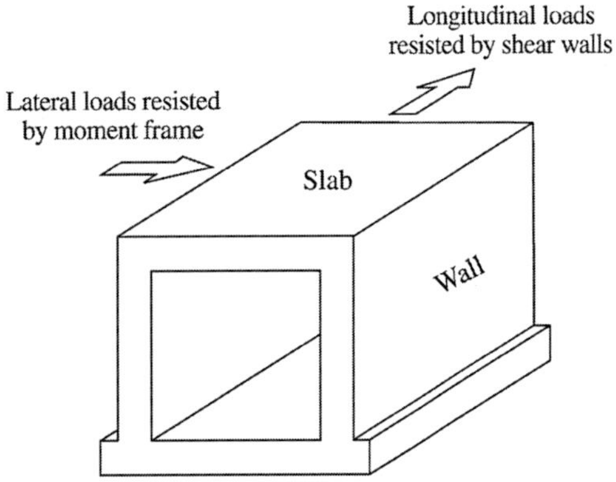

Figure C6-1. Idealized slab-wall moment frame.

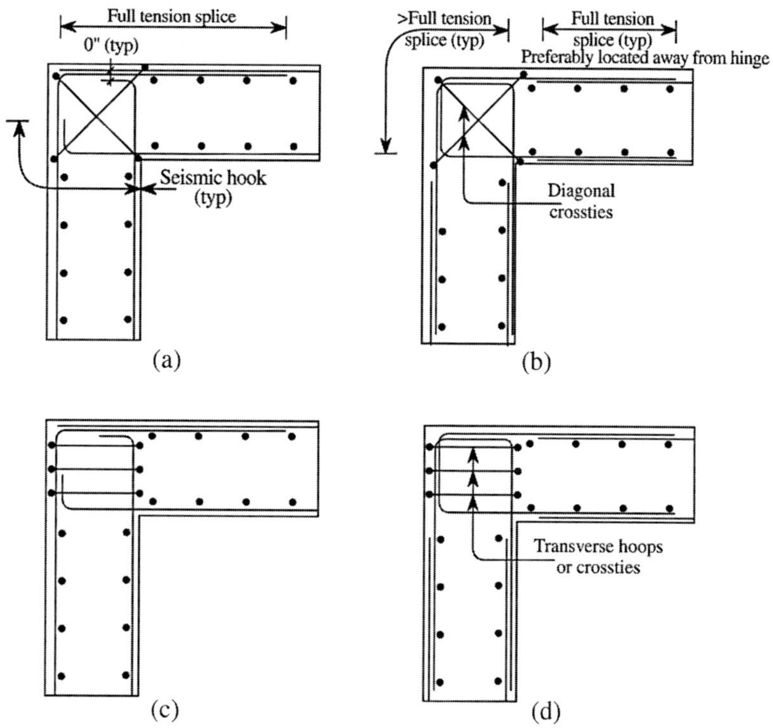

Figure C6-2. Typical joint details: (a) and (b), diagonal crossties; and (c) and (d), joint confinement.

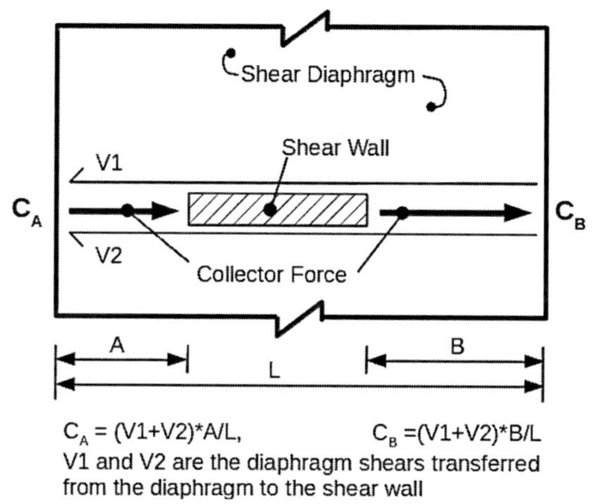

Figure C6-3. Collector element design forces.

and may lead to unacceptable seismic performance, particularly if a fraction of the reinforcing crossing the joint is discontinuous.

Often sequenced construction and/or pour strips can be used to control cracking, which eliminates the need for joints. If a joint with discontinuous reinforcing is necessary, separate seismic load paths should be supplied on each side of the joint and the relative motions, including the potential for pounding across the joint, should be evaluated.

C6.3 ANCHORAGE

Reliable anchorage is achieved by properly installing cast-in-place bolts, headed studs, or undercut-type expansion anchors, or by welding to embedded plates. Other types of expansion anchors are not recommended for vibratory environments (i.e., support of rotating machinery), for heavy equipment, or for sustained tensile loads.

Anchors that pass tests according to ACI 355.2 (mechanical anchors) (2007) or ACI 355.4 (adhesive anchors) (2011) are considered acceptable for application to nuclear structures.

Cast-in and postinstalled mechanical anchors that comply with the provisions of ACI 349 (2013b) for design and testing are considered acceptable for application to nuclear structures.

Provisions for adhesive anchors are not provided in either ACI 349 (2013b) or ACI 318 (2008), but they are available in ACI 318 (2014). The ACI 318 (2014) provisions require adhesive anchors to pass the simulated seismic testing of ACI 355.4, and installer certification and continuous inspections are required for adhesive anchors installed in horizontal or upwardly inclined orientations. Adhesive anchors that are designed, installed, and tested according to ACI 318 (2014) are considered acceptable for application to nuclear structures. The environmental qualification of adhesive anchors should address the effects of elevated temperature, radiation, moisture, sustained loading, and vibratory loading expected over the life of the anchorage.

CHAPTER C7
SPECIAL CONSIDERATIONS

This chapter provides evaluation procedures for rocking, overturning, and sliding demands calculated in accordance with ASCE 4 (2017b). As with other types of analysis, the analyst must determine and justify whether to treat the problem in one, two, or three dimensions, and whether the components of response can be decoupled or must be considered simultaneously.

C7.1 ROCKING AND SLIDING OF UNANCHORED BODIES

For multiple time-history inputs, each of which produces essentially the same 5% damped response spectrum, considerable variability has been found in the computed sliding distance, and moderate variability has been found in the computed rocking angle. Furthermore, if the coefficient of friction differs by as little as 0.1 between the plus and minus directions, sliding ratchets in the direction of the lower coefficient of friction. The capacity factor of safety of 2, defined in Section 7.1, was judged by the Dynamic Analysis of Nuclear Structures (DANS) Committee to produce design values of sliding and rocking that have less than about a 1% probability of being exceeded for the design input motion and less than about 10% probability of being exceeded for 150% of the design input motion.

C7.2 BUILDING SLIDING AND OVERTURNING

The approximate methods presented in ASCE 4 (2017b) are strictly only applicable for sliding and rocking of rigid bodies. However, these approaches may also be used to estimate the sliding and rocking of flexible buildings, as long as the fundamental building frequency, f_b, exceeds the effective sliding and rocking frequencies, f_{eS} and f_e, respectively. When f_b does not exceed the sliding frequency, f_{eS}, a conservative estimate of the sliding displacement in Section 11.3.1 of ASCE 4 (2017b) can be obtained by substituting f_b for f_{eS}. When f_b does not exceed the rocking frequency, f_e, then a conservative estimate of the rocking angle, θ_o, can be obtained from Section 11.3.2 of ASCE 4 (2017b) by substituting f_b for f_e.

Alternatively, an acceptable approach for analyzing buildings with foundation uplift is given in Yim and Chopra (1985).

Typically, static resistance is the resistance to overturning with respect to the corner of the foundation. For a rigid object on a rigid foundation, uplift occurs when the seismic load exceeds the static resistance. This uplift does not reduce the contact area because the contact area is a point.

For a near-rigid building on soil springs, the contact area starts to reduce as soon as there is tension on one side. It continues to reduce as the seismic load increases until there is a soil or building failure caused by the high-contact bearing reactions.

The foundations of most nuclear structures are embedded. The side traction against sliding may be reduced because of the separation between the soil and the structure. One method to account for the reduced traction is to assume no connectivity between the structure and the lateral soil over the upper half of the embedment or 20 ft (6 m), whichever is less. However, full traction between the structure and the side soil may be assumed if adjacent structures can produce appropriate surcharge or if a proper justification is provided. The extent of potential separation may also be evaluated by calculating the seismic and static soil pressure. These two methods are consistent with ASCE 4 (2017b).

For a deeply embedded structure as defined in Sections 7.2.1 and 7.2.2, sliding and/or overturning are not credible.

For structures that are highly sensitive to small displacements, wave passage effects and differential motions of soil layers may need to be considered.

C7.3 SEISMIC SEPARATION

Adjacent structures, systems, and components, denoted here as SSC_1 and SSC_2, are to be separated by a minimum distance to achieve the goals of (a) less than a 1% probability of impact in DBE shaking, and (b) less than a 10% probability of impact for 150% DBE shaking. Equation (7-5) uses the square root of the sum of the squares (SRSS) rule for combining the displacements of SSC_1 and SSC_2 along a common line. The factor of 2.0 is imposed on the SRSS displacement to achieve the two performance goals.

If SSC_1 and SSC_2 are not founded on a common foundation, the displacements of the two foundations should be included in the calculation of δ_1 and δ_2. The foundation contribution to the displacements should include components from both soil deformation and foundation flexibility, as appropriate. If performance goals are evaluated explicitly using nonlinear response history analyses, then the calculation of displacements δ_1 and δ_2 shall also account for inelastic action in any structures, systems, components, foundations, or soil supporting SSC_1 and SSC_2.

Section 7.3 requires that displacements based on the linear elastic response of short-period structures be further increased. It is well established that calculated inelastic displacements of short-period structures can substantially exceed the calculated elastic displacements of a corresponding system with the same elastic stiffness and mass. Coefficients in the archival literature, including greater than 1 Whittaker et al. (1998), may be used to estimate inelastic displacements and to increase δ_1 and δ_2 as determined by elastic analysis. Such increases are likely to be needed for structures, systems, and components assigned to Limit States A, B, and C. In lieu of calculations, increasing the calculated elastic displacement by a factor of 1.8 has been found to be conservative for most structural systems (Whittaker et al. 1998), and it therefore satisfies the requirements of Section 7.3 for any structural period.

C7.5 UNREINFORCED MASONRY WALLS USED FOR PARTITIONS, BARRIERS, OR RADIATION SHIELDING

Standard TMS 402/602 (TMS 2016) provides code requirements and specifications for masonry structures. The requirements of TMS 402/602 may need to be adjusted per Section 1.5, and Appendices A and B of this standard.

C7.6 PROVISIONS FOR CONSTRUCTION EFFECTS

Additional lateral wall pressures caused by soil compaction of fills adjacent to embedded walls can be estimated following the procedures described in Duncan and Seed (1986), Terzaghi et al. (1995), and Duncan et al. (1991). Dynamic lateral wall pressures are typically evaluated following the elastic procedures described in ASCE (2017b) and Ostadan and White (1998).

An evaluation of construction sequence (concrete pour sequence) to limit potential differential settlement effects is recommended.

CHAPTER C8
SEISMIC QUALIFICATION OF EQUIPMENT AND DISTRIBUTION SYSTEMS

The philosophy used to obtain the desired performance level according to the graded approach for equipment and distribution systems is described in this section. There are four steps to qualification of equipment or distribution systems:

1. Define the level of safety and the associated safety requirements. This is addressed in ANS 2.26 (ANS 2017).
2. Develop the seismic and nonseismic demand.
3. Determine the component capacity.
4. Compare the demand to capacity and document that the capacity is greater than the required demand.

Steps 2, 3, and 4 are addressed by this standard.

C8.2 QUALIFICATION BY ANALYSIS

The seismic motion at the support(s) of the distribution system, equipment, or component is based on an elastic analysis of the supporting structure for the design basis earthquake (DBE) that is defined by the Seismic Design Category (SDC) of the component. In this context, the *supporting structure* means the supporting structure, system, or equipment. This elastic seismic analysis is performed according to Chapter 3 of this standard, which refers to seismic analysis procedures of ASCE 4 (2017b). The ASME BPVC, Section III, Nonmandatory Appendix N (ASME 2019b), can also be used. For acceleration-based seismic input, the inelastic energy absorption factor (F_μ) of the supporting structure is set to 1.0 unless it can be justified, considering the potential structure overstrength, that the inelastic response of the supporting structure will effectively lower the support motion.

If the seismic input to a component is displacement based and the limit state of the supporting structure is other than Limit State D, then it may be nonconservative to assume that $F_\mu = 1.0$ for the analysis of the supporting structure. An example of this situation is a component that is supported at two elevations in the supporting structure. In this case, setting $F_\mu = 1.0$ in the analysis of the supporting structure could limit the differential displacements in the component and thus limit the seismic demand in the component.

The F_μ values in Table 8-1 are consistent with R_p values referenced in Table 13.6-1 of ASCE 7 (2017a) when it is understood that the allowable stresses for ASME Service Level D may be as high as twice the yield strength, whereas the ASCE 7 and other structural code allowables are typically not higher than the yield strength. The guidance of ASCE 4 (2017b), Chapter 6, "Input for Subsystem Analysis," can be used to define the input motion, including reduction of spectral amplitudes.

The application of equipment and distribution systems inelastic energy absorption factor, F_μ, provided in Table 8-1, is defined by the specified limit state and is only to be used for qualification by analysis. F_μ is not used when qualification is performed by test or experience data methods because the limit state is indirectly specified for the component by observed behavior (e.g., functional requirements and structural integrity) that defines the acceptance criteria.

The capacity of the component qualified by analysis is based on the specified code provisions and the allowable stress limits, such as those for various ASME service levels. For example, Service Level D in ASME BPVC, Section III (2019b) is a faulted condition, where the structures, systems, and components (SSCs) have gross deformation and require repair or replacement (i.e., similar to Limit State A or B of this standard). It generally includes analysis for internal pressure, deadweight, transients, and safe shutdown earthquake (SSE). SSE is somewhat equivalent to the DBE of this standard. Thus, the seismic design of a mechanical component whose failure is defined by Limit States A or B can follow ASME III BPVC, Section III (2019b) Service Level D stress limits, as has been permitted in Section 8.2.3.2 of this standard, but not if the SSC failure is defined by Limit State C or D.

For active components, the allowable stress levels are more conservative to ensure that the active functional requirements are met. For active components, Service Level A or B of ASME BPVC, Section III is often specified. Thus, active SSCs whose failure is defined by Limit States C or D can be designed with Service Level B or A stress limits, respectively.

C8.3 QUALIFICATION BY TESTING AND EXPERIENCE DATA

As nuclear facilities age, seismic qualification of replacement components can become a significant cost. To help reduce costs, the nuclear industry initiated the Seismic Qualification Reporting and Testing Standardization (SQURTS) program. The program was conceived in the early 1990s and allows program members to access a database of test specifications and test results, thereby reducing costs. The database of records extends from the present to the early 1990s. The program is managed by the Electric Power Research Institute (EPRI), and further information is available on the EPRI website (epri.com).

Experience-based approaches for verifying seismic adequacy of certain components have been successfully implemented for more than 20 years. The Seismic Qualification Utility Group (SQUG) generic implementation procedure (GIP) (EPRI 1991b; SQUG 2000, 2001; USNRC 1992, 1997) and the DOE seismic evaluation procedure (DOE 1997) are examples of these approaches. Engineers who apply these experience-based approaches are required to meet certain professional experience requirements and have formal training in the application of these methods. IEEE 344 (2013) and ASME QME-1 (2017a) permit the use of earthquake experience data under certain conditions related to the earthquake database and to the evaluation procedure.

Seismic interaction is the physical interaction of any structures, piping, or equipment with a nearby item of safe shutdown equipment caused by relative motions from an earthquake. The effect of seismic interaction of adjacent structures, systems, and components (SSCs) would be part of the final assurance of seismic qualification of the safety-related item. Interaction of nonsafety SSCs that could affect the safety function of equipment or distribution systems typically has the same sign design category (SDC) but may have a less stringent (lower performance) limit state. Additional information on seismic interaction can be found in the SQUG GIP.

C8.3.2 Demand for Qualification by Testing and Experience Data

C8.3.2.1 Demand for Qualification by Testing. The purpose of this section is to provide a basis for the value of γ_{test}, that is, the ratio of the test response spectrum (TRS) to the required response spectrum (RRS) that appears in Equation (8-3). The goal is to express the ratio of TRS to RRS in terms of quantities that are known or can be estimated.

Equation (C2-13) of this standard defines the minimum required median factor of safety ($F_{50\%R}$) to achieve the target performancer goal P_F when the hazard exceedance frequency H is defined by H_P equal to the performance goal P_F. From Equation (C2-13), the minimum required $F_{50\%R}$ is given in Table C2-2 as a function of A_R and β.

The design response spectrum (DRS) is defined by Equation (2-1) of this standard:

$$\text{DRS} = \text{SF} \times \text{UHRS}_{H_P} \qquad \text{(C8-1)}$$

where the scale factor (SF) is defined by Equation (2-3) of this standard and UHRS_{H_P} is the uniform hazard response spectrum (UHRS) defined at the hazard exceedance frequency H_P equal to the performance goal P_F. The seismic demand RRS is computed for the DRS (i.e., the scaled-down UHRS_{H_P}) and not for the UHRS_{H_P} itself. The required response spectrum UHRS_{H_P} corresponding to the UHRS_{H_P} is related to the RRS corresponding to the DRS in a manner analogous to Equation (C8-1):

$$\text{RRS} = \text{SF} \times \text{RRS}_{H_P} \text{or}\, \text{RRS}_{H_P} = (\text{RRS})/\text{SF} \qquad \text{(C8-2)}$$

Therefore, the achieved median factor of safety ($F_{50\%A}$) for equipment qualified by test is given by

$$\begin{aligned} F_{50\%A} &= (\text{TRS}/\text{RRS}_{H_P}) \times F_{R50\%} \times F_{C50\%} \\ &= (\text{TRS}/\text{RRS}) \times \text{SF} \times F_{R50\%} \times F_{C50\%} \qquad \text{(C8-3)} \end{aligned}$$

where

(TRS/RRS) = Ratio of the TRS to the RRS computed for the DRS;

SF = Scale factor used to define the RRS, as given by Equation (2-3) of this standard;

$F_{R50\%}$ = Estimate of the median response factor of conservatism in the computation of the RRS; and

$F_{C50\%}$ = Median estimate that any specific piece of equipment exceeds the TRS capacity of a similar tested piece of equipment.

To achieve the target performance goal, P_F, the achieved median factor of safety, $F_{A50\%}$, must equal or exceed the minimum required median factor of safety $F_{R50\%}$ defined in Table C2-2 of this standard.

Setting $F_{50\%A}$ equal to $F_{50\%R}$, the required (TRS/RRS) ratio from Equation (C8-3) becomes

$$(\text{TRS}/\text{RRS})_R \geq \frac{(F_{50\%R}/\text{SF})}{F_{R50\%} \times F_{C50\%}} \qquad \text{(C8-4)}$$

To solve Equation (C8-4) for $(\text{TRS/RRS})_R$, one must make estimates for $F_{R50\%}$ and its variability, β_R, and $F_{C50\%}$ and its variability, β_C.

Because the design response is intended to be at the 80% nonexceedance probability for the DRS:

$$F_{R50\%} = \exp[0.842\beta_R] \qquad \text{(C8-5)}$$

Salmon and Kennedy (1994) estimate a reasonable average response variability to be about

$$\beta_R = 0.35 \qquad \text{(C8-6)}$$

Thus,

$$F_{R50\%} = 1.343 \qquad \text{(C8-7)}$$

Salmon and Kennedy (1994) also estimate the test capacity from a qualification test to be in the 90% to 98% exceedance probability range based on the conservatism of the test. Taking a midpoint value of 95% exceedance probability,

$$F_{C50\%} = \exp[1.645\beta_C] \qquad \text{(C8-8)}$$

In addition, Salmon and Kennedy (1994) estimate a reasonable capacity variability to be

$$\beta_C = 0.20 \qquad \text{(C8-9)}$$

Table C8-1. (TRS/RRS) Ratios as a Function of Hazard Curve Slope Factor (A_R).

A_R	$F_{50\%R}$ $\beta = 0.40$	SF Eq. (C2-6)	(TRS/RRS)
1.50	1.58	0.667	1.27
1.75	1.39	0.571	1.30
2.00	1.30	0.522	1.33
2.25	1.26	0.510	1.32
2.50	1.22	0.500	1.31
2.75	1.20	0.490	1.31
3.00	1.18	0.482	1.31
3.25	1.17	0.474	1.32
3.50	1.16	0.467	1.33
3.75	1.15	0.461	1.34
4.00	1.14	0.455	1.34
4.25	1.14	0.450	1.36
4.50	1.13	0.450	1.35
4.75	1.13	0.450	1.35
5.00	1.12	0.450	1.33
5.25	1.12	0.450	1.33
5.50	1.11	0.450	1.32
5.75	1.11	0.450	1.32
6.00	1.11	0.450	1.32

Table C8-2. Median and Variability Factors for Specific Test Experience and Earthquake Experience Databases.

Database Capacity	$F_{R50\%}$	β_R	$F_{C50\%}$	β_C	β	γ
Nonrelay GERS	1.343	0.35	1.49	0.25	0.43	1.30
Relay GERS	1.343	0.35	1.07	0.20	0.40	1.75
SQUG reference spectrum	1.343	0.35	2.35	0.45	0.57	1.00

Table C8-3. γ Factors as a Function of Hazard Curve Slope Factor (A_R).

A_R	SF Eq. (C2-6)	$F_{50\%R}$ β = 0.43	γ	$F_{50\%R}$ β = 0.40	γ	$F_{50\%R}$ β = 0.57	γ
1.5	0.667	1.69	1.27	1.58	1.65	2.52	1.20
2.0	0.522	1.36	1.30	1.30	1.73	1.72	1.04
2.5	0.500	1.26	1.26	1.22	1.70	1.50	0.95
3.0	0.482	1.21	1.26	1.18	1.70	1.41	0.92
3.5	0.467	1.19	1.27	1.16	1.73	1.35	0.91
4.0	0.455	1.17	1.28	1.14	1.74	1.31	0.91
4.5	0.450	1.15	1.28	1.13	1.75	1.28	0.90
5.0	0.450	1.14	1.27	1.12	1.73	1.26	0.89
6.0	0.450	1.13	1.25	1.11	1.72	1.23	0.87

Thus,

$$F_{C50\%} = 1.390 \quad \text{(C8-10)}$$

The overall β is defined by

$$\beta = \sqrt{\beta_R^2 + \beta_C^2} \quad \text{(C8-11)}$$

For $\beta_R = 0.35$ and $\beta_C = 0.20$,

$$\beta = 0.40 \quad \text{(C8-12)}$$

Therefore, to be consistent with $F_{R50\%}$ from Equation (C8-7) and $F_{C50\%}$ from Equation (C8-10), $F_{50\%R}$ in Equation (C8-4) should be from the $\beta = 0.40$ column of Table C2-2. Based on $F_{R50\%}$ from Equation (C8-7) and $F_{C50\%}$ from Equation (C8-10), Equation (C8-4) gives

$$(\text{TRS/RRS})_R = (F_{50\%R}/\text{SF})/(1.343)(1.390) = (F_{50\%R}/\text{SF})/1.867 \quad \text{(C8-13)}$$

where $F_{50\%R}$ is for $\beta = 0.40$, and SF is from the last column (Equation [C2-6] column) of Table C2-4 in this standard. Table C8-1 tabulates the resultant (TRS/RRS) ratio from Equation (C8-13) for a range of A_R ratios. Note that the computed (TRS/RRS) ratio ranges from 1.27 to 1.36 with a median ratio of 1.32. Thus, the recommended (TRS/RRS) value is

$$(\text{TRS/RRS}) = (4/3) = 1.33 \quad \text{(C8-14)}$$

C8.3.2.2 Demand for Qualification by Test Experience Data and Earthquake Experience Data. The demand for qualification by test experience data and earthquake experience databases is given by Equation (8-3) for test experience data and Equation (8-4) for earthquake experience data, where the appropriate γ factor is required to achieve the target performance goal. The γ factor can be determined for either test experience data (γ_{TES}) or earthquake experience data (γ_{EED}) by substituting γ for $(\text{TRS/RRS})_R$ in Equation (C8-4):

$$\gamma \geq \left(\frac{F_{50\%R}/\text{SF}}{F_{R50\%} \times F_{C50\%}}\right) \quad \text{(C8-15)}$$

The median response factor $F_{R50\%}$ is given in Equation (C8-7), and its variability β_R is given in Equation (C8-6). The scale factor is defined by Equation (C2-6) in this standard, with specific values shown in Table C2-4 of this standard as a function of A_R. The required $F_{50\%\text{R}}$ is defined by Equation (C2-13) of this standard with specific values shown in Table C2-2 as a function of the hazard curve slope factor A_R and the composite (or overall) variability β, where β is given by Equation (C8-11). Thus, to solve Equation (C8-15) for γ for various test experience and earthquake experience databases, the estimated median capacity factor $F_{C50\%}$ and its variability β_C must be defined.

For the following specific test and earthquake experience databases, estimates of $F_{C50\%}$, β_C, and β are provided in Salmon and Kennedy (1994) and are summarized in Table 4 of that reference:

- Nonrelay generic equipment ruggedness spectra (GERS),
- Relay GERS, and
- SQUG reference spectrum.

These estimates are summarized herein in Table C8-2.

Table C8-3 tabulates the resulting γ factors from Equation (C8-15) for a range of A_R ratios for each of these specific databases. Based on results in Table C8-3, the recommended γ values are shown in Table C8-2.

For other databases, a similar approach can be used to define the γ factor required to meet the performance goals of this standard.

Additional information on GERS can be found in EPRI (1991a, 1996).

CHAPTER C9
SEISMICALLY ISOLATED STRUCTURES

C9.1 INTRODUCTION

Figure C9-1 identifies components of a seismically isolated nuclear structure. The isolator units (also termed *isolators* and *bearings*) are installed in a nearly horizontal plane beneath a basemat that supports the nuclear construction, which is defined as the *superstructure*. The isolators are installed atop pedestals and a foundation, which is defined as the *substructure*. The moat is a space in which the isolated superstructure can move without restriction in the event of earthquake shaking.

The implementation of a horizontal seismic isolation system at the base of a safety-related nuclear structure serves to substantially reduce the horizontal inertial forces on the superstructure and substructure, as indicated in Figure C9-2a. The increase in displacements shown in Figure C9-2b is accommodated by and large by the seismic isolators that are proportioned to support high axial loads at large lateral displacements, as required by Chapter 9.

Chapter 9 addresses horizontal isolation of nuclear structures only because no commercially available isolator can provide vertical isolation at this time.

Other hazards, including flooding of the basement, could potentially affect the response of the seismic isolation system to earthquake shaking. These hazards should be considered in the design and detailing of the nuclear structure so as to enable the isolation system to perform its safety function.

Table 12-1 of ASCE 4 (2017b) presents the minimum performance expectations and the associated acceptance criteria for the isolation system; superstructure framing; substructure framing; structures, systems, and components (SSCs); umbilical lines; and stops. (More stringent expectations may be imposed by regulatory agencies.) The performance expectations of Section 1.3 of this standard form the basis of this table and the provisions of Chapter 9, namely, (1) 1% probability of unacceptable performance for 100% design basis earthquake (DBE) shaking, and (2) 10% probability of unacceptable performance for 150% DBE shaking. Beyond design basis earthquake (BDBE) shaking is defined in Chapter 9 as 150% DBE shaking.

Four performance statements for achieving the two performance objectives of this standard were assumed in the writing of this chapter, namely,

1. Individual isolators should suffer no damage in DBE shaking.
2. The probability of the isolated nuclear structure impacting the surrounding structure or the moat wall for 100% (150%) DBE shaking should be 1% (10%) or less.
3. Individual isolators should sustain gravity and earthquake-induced axial loads at 90th percentile lateral displacements consistent with 150% DBE shaking.
4. The probability of unacceptable performance in the isolated superstructure for 100% (150%) DBE shaking should be 1% (10%) or less.

Performance statement 1 is realized by production testing of each isolator supplied to a project for the 80th percentile DBE displacement and coexisting gravity and earthquake-induced axial forces. Analysis can be used in support of performance statement 2, provided that the isolators are modeled correctly and the ground motion representations are reasonable. Huang et al. (2009, 2012) showed that 90th percentile displacement for 150% DBE shaking is greater than 99th percentile displacement for DBE shaking, and so the former is used to establish the minimum required clearance to the stop. Performance statement 3 can be realized by prototype testing of a limited number of isolators at a displacement equal to the clearance to the stop and coexisting axial forces, noting that an isolation system is composed of tens to hundreds of isolators and that failure of the isolation system would have to involve the simultaneous failure of a significant percentage of the isolators in the system. Performance statement 4 is checked by analysis of the capacity of the isolated superstructure at 100% DBE and BDBE shaking using procedures presented in Chapter 1 of this standard. If the probabilities of failure at DBE and BDBE shaking exceed the limits in Section 1.3 of this standard, the capacity (strength) of the isolated superstructure should be incremented until the performance statement is achieved. Adequate performance of the foundation is achieved by designing for forces delivered by the isolation system at horizontal displacements to the stops.

Seismic demands are calculated for DBE and BDBE shaking. Results of DBE analysis are used for (a) calculating design loads on the superstructure, (b) generating in-structure response spectra for design of SSCs, and (c) establishing displacements for production testing of isolators. Design loads on the isolated superstructure may be increased if the performance expectations of this standard are not met, as noted in the preceding paragraph. Results of BDBE (150% DBE) analysis are used to (a) select the required clearance to a physical stop or displacement restraint, and (b) establish displacements and forces for prototype testing of isolators.

A stop, which may be a moat wall, is used to prevent excessive displacement of the isolation system and remove the isolation system from accident sequences involving earthquake shaking. The clearance to the stop along each horizontal axis used to define the seismic input shall equal or exceed the 90th percentile BDBE displacement along that axis. The clearance to stop (CS) is the 90th percentile BDBE displacement, which exceeds the clearances to the stop along the horizontal axes used to define the seismic input, but is not greater than the vector sum of the two

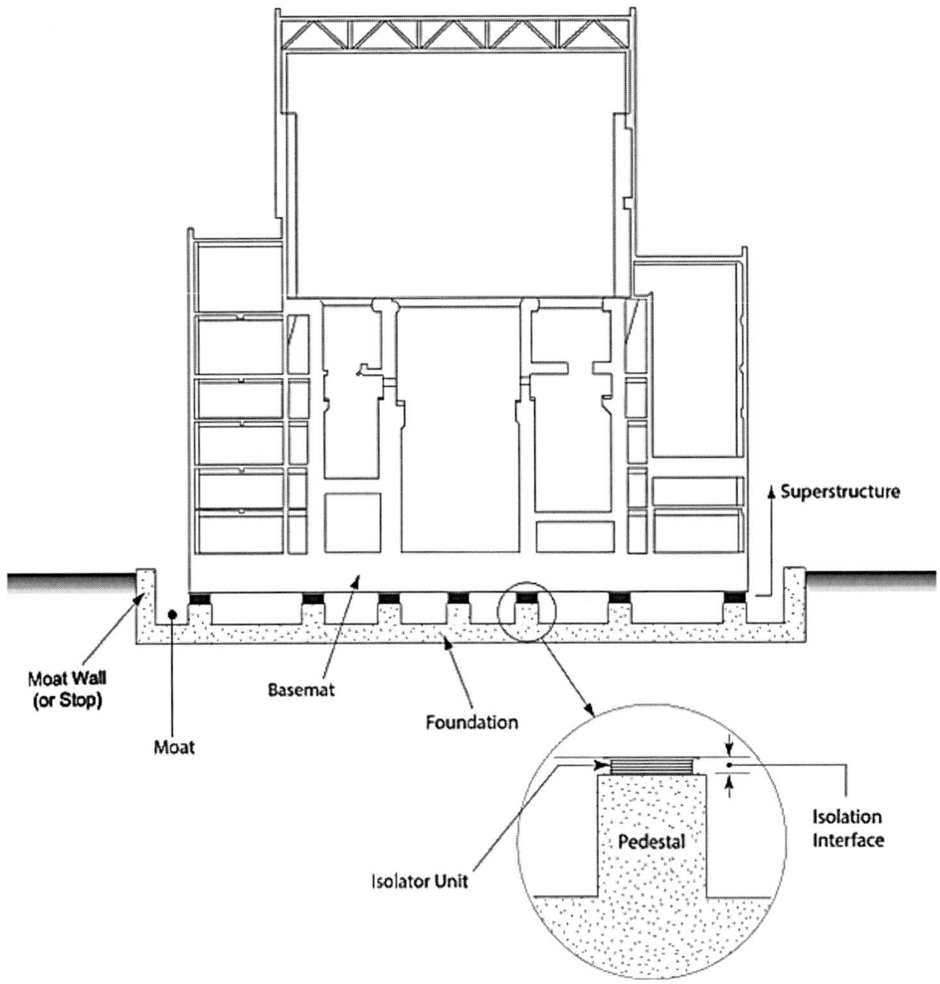

Figure C9-1. Isolated nuclear structure.

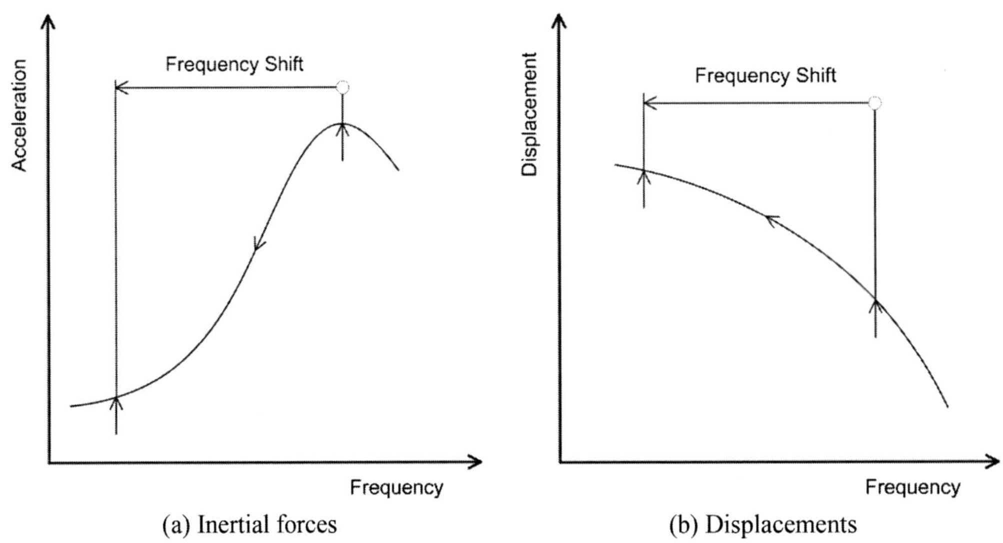

Figure C9-2. Effect of seismic isolation on inertial forces and displacements in structures.

clearances. The displacement CS is used for prototype testing of isolators.

Three types of isolators have been qualified for use in safety-related nuclear structures: low-damping (natural) rubber (LDR) isolators, lead–(natural) rubber (LR) isolators, and friction pendulum (FP) sliding isolators. Each has been tested extensively, can be modeled for nonlinear response-history analysis, and has been deployed in mission-critical structures.

Low-damping (LD) elastomeric bearings are composed of alternating layers of natural rubber and steel, and they can be modeled as viscoelastic components. The shear modulus of the rubber ranges between 60 lb/in.^{2} (0.41 MPa) and 120 lb/in.^{2} (0.83 MPa). The equivalent viscous damping is between 2% and 4% of critical.

Lead–rubber (LR) elastomeric bearings are constructed similarly to low-damping rubber bearings but include a central lead core to dissipate earthquake-induced energy. The hysteresis loop for the LR bearing is bilinear according to Figure C9-3 and is defined by a zero-displacement force intercept, $Q_d W$, an elastic stiffness, K_u, and a second-slope stiffness, K_d, where W is the supported weight.

Sliding bearings with restoring force provided by gravity also have the hysteresis loop of Figure C9-3. In the friction pendulum (FP) family of bearings, the second-slope stiffness is equal to the supported weight divided by the radius of curvature of the sliding surface.

Other types of bearings may be acceptable for safety-related nuclear applications in the United States later, but the five tasks of Section 9.1 should be performed before a new type of isolator can be deployed in a safety-related nuclear structure. Accelerated aging tests should not be used in task 3 to characterize the evolution of the mechanical properties of elastomeric bearings (Constantinou et al. 2007).

Isolation systems involving more than one type of isolator (e.g., elastomeric and sliding) are not permitted because of incompatibility of vertical displacements: Elastomeric bearings shorten when displaced laterally, whereas sliding bearings either retain their height or increase in height when displaced laterally.

The design of a seismically isolated nuclear structure is performed using the results of dynamic analysis. Static analysis methods are not permitted for the calculation of isolator response. Chapter 12 of ASCE 4-16 (2017b) describes three acceptable methods of dynamic analysis. All three methods involve soil–structure interaction analysis.

Diaphragms are required above and below the isolation system. In this section, the diaphragms above and below the isolation system are identified as basemat and foundation, respectively.

Constantinou et al. (2007) and Naeim and Kelly (1999) provide information on LD, LR, and FP seismic isolators. Huang et al. (2008, 2009, 2010) provide data and information in support of the isolation of nuclear structures and the provisions of this chapter. Constantinou et al. (1999, 2007), Thompson et al. (2000), and Morgan et al. (2001) provide the data used by Huang et al. (2009, 2012) to characterize the impact of variations in isolator material properties on the displacement response of seismic isolation systems. Kumar et al. (2014, 2015b) provide an open-source model of LD and LR bearings suitable for nonlinear response-history analysis of elastomeric isolation systems under extreme earthquake loadings. Kumar et al. (2015a) provide the corresponding open-source model of the FP bearing.

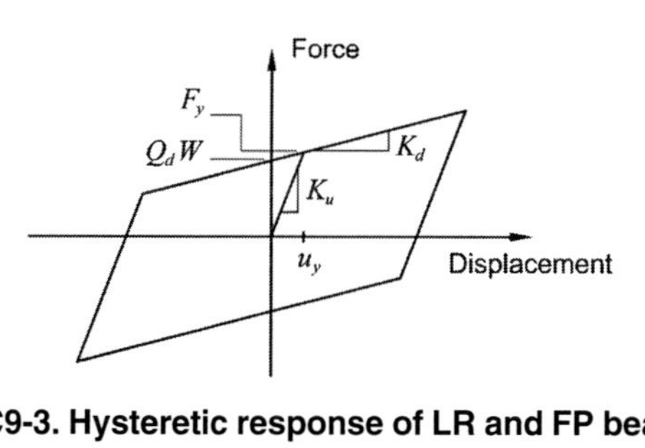

Figure C9-3. Hysteretic response of LR and FP bearings in a horizontal direction.

The mandatory language and commentary to this chapter are based in part on provisions and commentary developed for the seismic isolation of buildings (ASCE 2010) and bridges (AASHTO 2010). The reader is referred to these standards and guidelines for information not presented here.

C9.2 GENERAL REQUIREMENTS

C9.2.1 Isolation System

C9.2.1.2 Vertical and Horizontal Load Resistance. The vertical loads used to check the adequacy of an isolator should be calculated using the maximum positive and minimum negative values of (a) dead load, (b) live load, and (c) earthquake loads associated with the DBE displacement, D_D, and the CS.

No permanent damage to the isolators is permitted for DBE displacements and forces. Short-term changes in hysteretic properties caused by energy dissipation (e.g., Kalpakidis et al. 2010, Kumar et al. 2014) are acceptable, provided it is demonstrated that the pre-earthquake mechanical properties are recovered quickly. Isolator damage is permitted for BDBE shaking, provided that it does not compromise the required vertical load resistance of the isolation system.

Each component of the isolation system, with the exception of the isolators, should resist the forces associated with isolator displacement CS in the elastic range to ensure that the isolation system can perform its intended function.

C9.2.1.3 Minimum Lateral Restoring Force. The restoring force requirement is intended to limit residual displacement as a result of an earthquake so that the isolated structure can survive aftershocks and future earthquakes. The rule is identical to that in ASCE 7 (2010) for isolated buildings and other structures.

C9.2.1.4 Wind Loads. Lateral forces in seismic isolators under design wind loads should not result in damage. The limits of Section 9.2.1.4 achieve this goal. The yield displacement for LR and FP bearings, u_y, is identified in Figure C9-3.

C9.2.1.5 Operating Conditions. The isolation system should be maintained in a temperature-controlled environment. The mechanical properties of LDR, LR, and FP isolation systems are not dependent on temperature in the range of 40 °F (4.4 °C) to 80 °F (26.67 °C). A greater range of operating temperatures may be acceptable if adequate performance is demonstrated by analysis.

C9.2.1.6 Inspection and Replacement. Although isolators should likely not need to be replaced after DBE or BDBE shaking, the design of the structural framing system should provide for access to the isolators for periodic and post-earthquake inspection and possible replacement.

The provisions require periodic inspection of isolators because seismic isolation has not been used for nuclear structures in the United States at the time of this writing. This requirement may be relaxed after experience is gained with isolation of nuclear structures.

The isolators, isolation system, framing above and below the isolators, and secondary systems crossing the isolation interface should be inspected after significant earthquake shaking. Components that are damaged should be replaced or repaired if the use of the facility is to be maintained.

Spare isolators of each type used in the isolation system should be stored on site and tested at 10-year intervals according to Section 9.5.3 to build further confidence in the technology among structural engineers implementing seismic isolation in nuclear structures.

C9.2.2 Isolators

C9.2.2.1 Mechanical Properties. Analysis was performed by Huang et al. (2009, 2012) to estimate displacements for LD, LR, and FP bearing isolation systems at representative sites across the United States for DBE and 150% DBE shaking. These analyses considered variability in both the seismic input (i.e., a distribution in spectral demand; see Huang et al. 2009, 2012) and the mechanical properties of the isolation system. Factors were developed and used in this standard to relate the mean displacement in the isolation system in DBE shaking to a displacement with 1% probability of exceedance in DBE shaking and a 10% probability of exceedance in 150% DBE shaking. The greatest variability considered for isolator properties was ±20% from the best-estimate values with 95% probability.

C9.2.2.2 Quality Assurance. Consistent with other critical components in nuclear structures, isolators shall be designed and tested to quality standards accepted across the nuclear industry, including ASME NQA-1 (ASME 2017b).

C9.2.3 Basemat and Foundation Designs. The plan dimensions of foundations for nuclear islands are generally large. Vertical deflection of the basemat and/or settlement of the foundation below a small part of the isolation system may lead to a redistribution of gravity loads to the isolators, which is generally undesirable. The basemat and the foundation should be designed and detailed to enable adjustment of the isolators in the event of settlement.

To provide redundancy in the gravity load-resisting system, the basemat should be designed to support gravity loads assuming the loss of a single isolator and to enable adjustments to the isolators in the event of settlement. Multiple calculations, assuming the loss of a different isolator in the system, should be performed to demonstrate the adequacy of the basemat.

C9.3 DISPLACEMENTS AND FORCES FOR DESIGN

C9.3.1 General. Huang et al. (2009, 2012) showed that displacement demands (and thus transmitted forces) associated with 10% probability of exceedance in 150% DBE shaking are greater than those associated with 1% probability of exceedance in DBE shaking. Accordingly, the performance expectations of Section 1.3 of this standard are achieved in this chapter using 90th percentile displacements and forces for 150% DBE (i.e., BDBE) shaking.

The 80th percentile DBE horizontal displacement, D_D, and the mean DBE horizontal displacement should be calculated by response-history analysis.

Design basis earthquake forces at the 80th percentile level should be calculated by response-history analysis.

The 90th percentile BDBE horizontal displacement, D_{BD}, can be calculated either directly by response-history analysis or by multiplying the mean DBE displacements by a factor of 3.0. The factor of 3.0 is based on the studies of Huang et al. (2009, 2012) and is a conservative value calculated for alternate nuclear power plant sites in the United States and different types of LR, LDR, and FP isolators. The factor of 3.0 should not be used for other types of isolators.

BDBE forces on the isolation system at the 90th percentile level can be calculated by multiplying the mean DBE forces by the ratio of the horizontal resistance of the isolation system at displacement D_{BD} to the horizontal resistance at the mean DBE displacement.

C9.3.2 Seismic Isolators. The displacement D_D is used for production testing of isolators.

The displacement D_{BD} is used to establish the displacement CS and thus the displacement for prototype testing of isolators.

The connections of the isolators to the substructure and superstructure shall be designed to resist the forces delivered by the isolation system at a displacement equal to CS to ensure that nonlinear response is confined to the isolators and that the isolation system does not form part of any accident sequence. The load factors and strength reduction factors presented in materials standards shall be used for design.

C9.3.3 Stop. A stop, which may be a moat wall, is used to prevent excessive displacement of the isolation system in the event of beyond design basis earthquake shaking and to eliminate isolator failure caused by earthquake shaking from any accident sequence.

Displacements along two orthogonal horizontal axes should be calculated at the center of mass of the isolated superstructure, projected down to the basemat. The axes (x, y) should align with those used to define the horizontal seismic input. Horizontal displacements at the 90th percentile level along each axis should be calculated. The clearance between the isolated superstructure and the stop along each axis shall be no less than the corresponding 90th percentile displacement.

The 90th percentile horizontal displacement in all directions, D_{BD}, exceeds the horizontal displacements at the 90th percentile level along the axes used to prescribe the seismic input, but it is less than the square root of the sum of the squares of the two displacements.

The CS is the displacement with a 10% or less probability of exceedance for BDBE shaking. The value is defined by the maximum possible horizontal displacement of the isolated superstructure within the stop or moat.

The stop should be designed for impact by the isolated superstructure. The velocity at impact may be calculated by either analysis for BDBE ground motions or by assuming cyclic response of the isolated superstructure to the 95th percentile BDBE displacement at a frequency equal to that calculated for the isolation system, assuming the best-estimate second-slope stiffness (K_d in Figure C9-3 for LR and FP isolation systems) and the effective seismic weight of the superstructure.

C9.3.4 Structures, Systems, and Components Above the Isolation Interface. The design of the superstructure, systems, and components above the isolation interface should be based on the results of response-history analysis for DBE shaking at the 80th percentile level.

Equivalent static methods of analysis according to Chapter 4 of ASCE 4 (2017b) can be used for the design of isolated components provided that (a) the fundamental horizontal frequency of the isolated component is more than five times greater than the horizontal frequency of the isolation system, and (b) the vertical frequency of the isolated component is greater than 50 Hz.

Calculations should be performed to demonstrate that the performance expectations of Section 1.3 are achieved. If the failure probabilities exceed either 1% for DBE shaking or 10% for BDBE shaking, the design forces should be increased to reduce the failure probabilities to the limiting values.

C9.3.5 Structures Below the Isolation Interface. To ensure that nonlinear response is confined to the isolators, the design of the foundation should be based on the forces delivered by the isolators at an isolation-system displacement equal to CS. The load factors and strength reduction factors presented in materials standards such as ACI 349 (ACI 2013b) shall be used for design.

C9.3.6 Systems and Components Crossing the Isolation Interface. Umbilical lines are systems and components that cross the isolation interface and may serve a safety-related function. The probability of failure of these umbilical lines should be less than 10% in BDBE shaking (and also 1% in DBE shaking, which should not control, as noted in a preceding section). The umbilical lines should be capable of sustaining a displacement equal to CS and should remain functional, with 90% confidence. Numerical simulations, full-scale dynamic testing, or a combination thereof, may be used to demonstrate adequacy.

C9.4 PEER REVIEW

A peer review of the design of an isolated nuclear structure should follow the requirements of Section 1.2.3 of ASCE 4 (2017b) and Section 9.1 of this standard for nonisolated structures and should also address those aspects of the design and specification related to the seismic isolators, isolation system, and SSCs that cross the isolation interface.

The isolation system peer review should include but not be limited to the following:

1. Review of numerical models of isolators;
2. Review of the soil–structure interaction (SSI) analysis and the resulting in-structure response spectra (according to ASCE 4);
3. Review of displacement and force calculations for the isolator units and all associated structures, systems, and components;
4. Review of the design of the umbilical lines;
5. Review of the design of the stop;
6. Review of the seismic monitoring program;
7. Review of the prototype test program;
8. Review of the production (quality control) test program;
9. Review of the isolator inspection and post-installation testing program; and
10. Review of postearthquake inspection protocols.

Review of the design of the isolation system and design review of the isolator-testing program is mandated by this standard because the consequences of isolator failure could be catastrophic, and isolator design and fabrication technology may be unfamiliar to many design professionals.

This standard requires that a team of registered design professionals perform the design review. The peer review team should be independent of the design team and other project contractors. The review team should include individuals with special expertise in one or more aspects of the design, analysis, testing, and implementation of seismic isolation systems.

C9.5 TESTING OF PROTOTYPE AND PRODUCTION ISOLATORS

C9.5.1 General. This section of the standard describes the two types of tests to be performed on seismic isolators: prototype and production.

Prototype bearing tests are performed on a small number of isolators of each type intended for use in the isolation system. These tests are conducted before isolator production for the project. The purpose of the tests is to ensure that the isolators perform as anticipated for DBE and BDBE shaking.

Production (or quality control) tests are performed on each isolator fabricated for a project. The production-testing program is less onerous than the prototype-testing program: Dynamic tests are not required, and the test displacement is that computed for the 80th percentile DBE displacement.

The prototype and production tests focus on horizontal seismic response of isolators. Tests may be required to confirm the mechanical properties of an isolator in the vertical (axial) direction.

C9.5.2 Prototype Testing

C9.5.2.1 Test Specimens. Three isolators of each type and size in the isolation system are fabricated for prototype testing. This requirement is identical to that for building structures.

C9.5.2.2 Record. The minimum data to be recorded in a prototype test should enable the calculation of axial force, axial displacement, horizontal force, and horizontal displacement histories, and the generation for force-displacement hysteresis loops.

A sampling rate of 200 Hz is sufficient to resolve the response of isolators that are tested at frequencies between 0.5 Hz and 0.01 Hz.

C9.5.2.3 Sequence and Cycles. Section 9.5.2.3 specifies a minimum sequence of tests for prototype isolators that exhibit either linear response or bilinear response according to Figure C9-3. Other tests may be specified on a project-specific basis, including tests to compute vertical stiffness. ASCE 7 (2010) and the AASHTO *Guide Specification for Seismic Isolation Design* (AASHTO 2010) include tests for isolators in buildings and bridges, respectively.

The number of cycles of loading for tests 2 and 3 is conservatively based on the data of Warn and Whittaker (2004).

The CS is computed at the 90th percentile level for 150% DBE shaking and is used for test 3.

C9.5.2.4 Test Specimen Adequacy. The acceptance criteria used for prototype bearing tests should be consistent with the assumptions made for the design of the seismic isolators and seismic isolation system, and they should therefore be developed on a project-by-project basis. Basic acceptance criteria for the three prototype tests are provided in this section for inclusion in the prototype isolator acceptance criteria developed by the responsible design professional.

Sample criteria can be found in ASCE 7 (2017a), and the AASHTO *Guide Specification for Seismic Isolation Des.* (2010).

C9.5.3 Production Testing

C9.5.3.1 Test Specimens. Each production isolator should be tested to ensure the use of high-quality isolators.

C9.5.3.2 Record. See Section C9.5.2.2.

C9.5.3.3 Sequence and Cycles. Project-specific acceptance criteria are generally developed for the production testing of isolators. For each type and size of bearing, shear force-lateral displacement relationships should range between limits specified by the responsible design professional. A minimum vertical stiffness for each type and size of isolator should be specified.

The reader is referred to ASCE 7 and the AASHTO *Guide Specification for Seismic Isolation Design* for additional information.

C9.5.3.4 Test Specimen Adequacy. The acceptance criteria used for production isolator tests should be consistent with the assumptions made for the design of the seismic isolators and seismic isolation system, and they are therefore developed on a project-by-project basis. The acceptance criteria provided in this section should be included in the production isolator acceptance criteria developed by the responsible design professional.

CHAPTER C10
QUALITY ASSURANCE PROVISIONS

The seismic analysis and design of nuclear facilities specified in this standard may be performed under the purview of the US Department of Energy (DOE) or the US Nuclear Regulatory Commission (NRC). The DOE and NRC have regulatory quality assurance (QA) requirements that are applicable throughout the planning and performance of design activities, including seismic analysis. The QA program can be composed of subprograms addressing the following elements:

1. Design verification and independent peer review;
2. Construction verification (including structural observation, inspection, and testing);
3. Quality assurance program; and
4. Software quality assurance.

C10.1 DESIGN VERIFICATION AND INDEPENDENT PEER REVIEW

ASME (2015b) includes detailed design verification processes that could be used for complying with the QA portion of this standard. DOE guidance from DOE (2013) is also appropriate for this purpose. The independent peer review prescribed for Seismic Design Category (SDC) 3, 4, and 5 structures, systems, and components (SSCs) should include the geophysical and geotechnical design basis and the seismic analysis and design of the SSCs. Because SDC 2 can cover a wide spectrum of SSCs, the extent of the seismic peer review, as determined by the design organization, should be commensurate with the severity of the failure consequence, hazard mitigation function, and design complexity of the SSCs.

C10.2 STRUCTURAL OBSERVATION, INSPECTION, AND TESTING

The graded approach for construction quality should be developed based on the SDC assigned. The minimum quality for SDC 2 and SDC 3 should meet the requirements of the *International Building Code* (ICC 2015), Chapter 17, with higher quality prescribed as needed using ASME (2015b). Additional guidance on developing inspection requirements can be obtained from ASCE (2000). The testing and inspection requirements of the applicable material code such as ACI (2013b), ACI (2014), AISC (2012), or other, should be specified.

C10.3 QUALITY ASSURANCE PROGRAM

The DOE QA requirements for nuclear facilities are defined in GPO (2011). These requirements are further elaborated in DOE directives and standards, including the use of a graded approach. The NRC QA requirements for nuclear power plants are established in USNRC (2008). In developing the details of the QA program to meet the DOE or NRC requirements, refer to their respective guidance documents. Nuclear facility design and seismic analysis detailed guidance is also available in DOE and NRC guidance documents.

C10.4 SOFTWARE QUALITY ASSURANCE

ASME (2015b), Part I and Subpart 2.7, provide requirements for quality assurance of software, either developed or acquired. Computer programs that are controlled and have already been verified and validated do not need such verification and validation (V&V) to be repeated before every execution of the computer program. Detailed requirements of the dedication activity, referred to as commercial grade dedication, are found in ASME NQA-1 (2015b). The dedication activity involves providing reasonable assurance that a commercial grade computer program or software service meets functional requirements, critical characteristics, and established acceptance criteria.

COMMENTARY TO APPENDIX A

ALTERNATE METHOD TO MEET ASCE 43 PERFORMANCE GOALS WHEN SEISMIC CAPABILITIES ARE DEFINED AT THE 50% PROBABILITY OF FAILURE LEVEL

CA3. SPECIFICATION OF MINIMUM FACTOR OF SAFETY TO BE APPLIED TO MEDIAN PROBABILITY OF FAILURE CAPABILITY

Equation (A-2) for the minimum required factor of safety is derived in Equation (C2-13). Table CA-1 tabulates $F_{50\%}$ as a function of A_R and β.

Note that in Equation (A-1), $F_{50\%}$ is applied to the $UHRS_{H_P}$ to define DRS_A.

Table CA-1. Required Safety Factor $F_{50\%}$ as a Function of A_R and β.

	$F_{50\%}$			
A_R	β = 0.30	β = 0.40	β = 0.50	β = 0.60
1.5	1.29	1.58	2.03	2.78
1.75	1.20	1.39	1.67	2.10
2.0	1.16	1.30	1.51	1.82
2.25	1.14	1.26	1.43	1.67
2.5	1.12	1.22	1.37	1.57
2.75	1.11	1.20	1.33	1.51
3.0	1.10	1.18	1.30	1.46
3.25	1.09	1.17	1.28	1.42
3.5	1.09	1.16	1.26	1.39
3.75	1.08	1.15	1.24	1.37
4.0	1.08	1.14	1.23	1.35
4.25	1.07	1.14	1.22	1.33
4.5	1.07	1.13	1.21	1.32
4.75	1.07	1.13	1.20	1.30
5.0	1.07	1.12	1.20	1.29
5.25	1.06	1.12	1.19	1.28
5.5	1.06	1.11	1.18	1.28
5.75	1.06	1.11	1.18	1.27
6.0	1.06	1.11	1.17	1.26

COMMENTARY TO APPENDIX B

ALTERNATE METHOD TO MEET ASCE 43 PERFORMANCE GOALS WHEN SEISMIC CAPABILITIES ARE DEFINED AT THE 10% PROBABILITY OF FAILURE LEVEL

CB3. SPECIFICATION OF MINIMUM FACTOR OF SAFETY TO BE APPLIED TO 10% PROBABILITY OF FAILURE CAPABILITY

When H is defined by $H_P = P_F$, the required median factor of safety, $F_{50\%}$, to achieve the performance goal, P_F, is derived thus:

$$F_{50\%} = \exp\left[\frac{1}{2}K_H\beta^2\right] \quad \text{(CB-1)}$$

This equation for $F_{50\%}$ is derived in Commentary, Section C2.2. Note that

$$F_{50\%} = \frac{C_{50\%}}{C_{H_P}} \quad \text{(CB-2)}$$

where C_{H_P} is the uniform hazard response spectra (UHRS) ground motion level that corresponds to this reference exceedance frequency H_P from the seismic hazard curve, and $C_{50\%}$ is the median capacity. Defining the required 10% failure probabilistic factor of safety, $F_{10\%}$, in terms of the DRS ground motion, C_{DRS}, by

$$F_{10\%} = \frac{C_{10\%}}{C_{\text{DRS}}} \quad \text{(CB-3)}$$

in which

$$C_{10\%} = C_{50\%}\exp(-1.282\beta) \quad \text{(CB-4)}$$

$$C_{\text{DRS}} = \text{SF} \times C_{H_P} \quad \text{(CB-5)}$$

where β is the natural logarithmic standard deviation of the fragility, −1.282 is the standard normal variate corresponding to the 10% nonexceedance probability, and SF is the scale factor defined by Equation (2-3) of this standard, which is repeated here as

$$\text{SF} = \text{Maximum}\ [\text{SF}_1;\ \text{SF}_2;\ \text{SF}_3] \quad \text{(CB-6a)}$$

$$\text{SF}_1 = A_R^{-1.0} \quad \text{(CB-6b)}$$

$$\text{SF}_2 = 0.6(A_R)^{-0.2} \quad \text{(CB-6c)}$$

$$\text{SF}_3 = 0.45 \quad \text{(CB-6d)}$$

From Equations (CB-2), (CB-3), (CB-4), and (CB-5),

$$F_{10\%} = F_{50\%}(\text{SF})^{-1}\exp(-1.282\beta) \quad \text{(CB-7)}$$

Table CB3-1. Required Safety Factor $F_{10\%}$ as a Function of A_R and β.

			$F_{10\%}$			
A_R	K_H	SF	$\beta = 0.30$	$\beta = 0.40$	$\beta = 0.50$	$\beta = 0.60$
1.5	5.68	0.667	1.32	1.41	1.61	1.93
1.75	4.11	0.571	1.43	1.46	1.54	1.70
2.0	3.32	0.522	1.51	1.50	1.53	1.61
2.25	2.84	0.510	1.52	1.47	1.47	1.51
2.5	2.51	0.500	1.53	1.47	1.46	1.46
2.75	2.28	0.490	1.54	1.47	1.43	1.42
3.0	2.10	0.482	1.55	1.47	1.42	1.40
3.25	1.95	0.474	1.57	1.48	1.42	1.39
3.5	1.84	0.467	1.58	1.49	1.42	1.38
3.75	1.74	0.461	1.60	1.49	1.42	1.38
4.0	1.66	0.455	1.61	1.50	1.43	1.37
4.25	1.59	0.45	1.63	1.51	1.43	1.37
4.5	1.53	0.45	1.62	1.50	1.42	1.36
4.75	1.48	0.45	1.62	1.50	1.41	1.34
5.0	1.43	0.45	1.61	1.49	1.40	1.33
5.25	1.39	0.45	1.61	1.49	1.39	1.32
5.5	1.35	0.45	1.61	1.48	1.39	1.31
5.75	1.32	0.45	1.61	1.48	1.38	1.31
6.0	1.29	0.45	1.60	1.47	1.37	1.30

Thus, from Equation (CB-1),

$$F_{10\%} = (\text{SF})^{-1}\exp\left[-1.282\beta + \frac{1}{2}K_H\beta^2\right] \quad \text{(CB-8)}$$

Table CB-1 tabulates $F_{10\%}$ as a function of A_R and β. The safety factor $F_{10\%}$ can be reasonably approximated by

$$F_{10\%} \approx 1.50 \quad \text{(CB-9)}$$

Note that in Equation (B-1), the factor $F_{10\%}$ is applied to the DRS from Section 2.2, whereas in Equation (A-1) of Appendix A, the factor $F_{50\%}$ is applied to the UHRS_{H_P}. Lastly, note that

$$\text{DRS} = \text{SF} \times \text{UHRS}_{H_P} \quad \text{(CB-10)}$$

where SF ranges between 0.667 and 0.45. Thus, $F_{10\%} \times \text{SF}$ is less than or equal to unity, whereas $F_{50\%}$ is greater than unity.

INDEX